Choosing the sex of your baby ...
made simple,
made certain.

iUniverse
Bloomington

Dr. Paul Gouda

C.Chem., P.R.M.D., Ph.D.

"Finally, a truly effective, easy to follow material, illustratively formatted for the layman and is supported by solid scholastic qualifications in the field of pharmaceutical - medical chemistry, along with real-life practical experience."

Dr. Andrew A. Hilam, Director, Optimum Health.

Choosing the sex of your baby-

or puppy.
Yes you can. With certainty.

An illustrated and simple manual for mammal applications, both human conception, and animal breeding.

Bloomington

Choosing the sex of your baby, made simple.

ISBN: 978-1-4759-3662-9 (sc)
ISBN: 978-1-4759-3663-6 (ebk)

Printed in the United States of America.
Iuniverse rev. date: 07/11/2012

Canadian Intellectual Property Office.
Certificate of registration of copyright number: **1096166**
Paul Gouda, Owner & Author. 06/06/2012
Issued pursuant to sections 49 & 53 of the Copyright Act.

Library and Archives Canada Cataloguing in Publication
Gouda, Paul
Choosing the sex of your baby-- made simple / Paul Gouda.
Includes index.
1. Sex pre-selection. 2. Reproduction. 3. Animal breeding.
I. Title.
QP279.G69 2012 613.9'4 C2012-903590-4

American Library of Congress Control Number: 2012912189

Dr. Paul R. Gouda's research papers in the arena of analytical - pharmaceutical chemistry have earned him the "elite scientist" status as a renowned chemist. Two doctorates and three decades of laboratory and pharmaceutical – medical research promise to make this unique paper, written in plain English for everyone's use, a great contribution to the community at large.

Dr. Samir Walid, MD., Ph.D.

For the past fifteen years I have run into many couples, friends of the author – some of whom specifically wanted a girl, and a few others who specifically wanted a boy – most of whom are now close friends of mine.
Each and every one of them – under the guidance of Dr. Gouda, a mutual friend, has succeeded in securing the desired gender of their baby.

Last year, in a casual conversation, I mentioned that I wanted to breed my "expensive" German Sheppard, an elite line of breeding from which the local police has a duty dog.
I commented that several friends wanted a pup, but they all wanted a male! I asked my friend Paul, the author, for his advice. He put the male and the female on a special supplementary diet that he provided, and he performed the artificial insemination after performing a specific semen treatment. Two months later, a litter of 11 pups, 11 males.
In fact, for the past 8 years it never ceased to amaze me how Dr. Gouda, who is a Great Dane lover, bred his Danes with a bold online announcement months prior to the whelping of the puppies that they "all" will be males – or they "all" will be females! And yes, he was never wrong.

Two years ago, coaching a Chinese couple who are mutual friends, Dr. Gouda published a congratulatory ad in the metropolitan newspaper congratulating them in advance on their yet-to-be conceived son. 11 months later, they named him "Paul." This book is an invaluable tool for every couple.

Dr. Timothy Bucha, Ph.D.

To order additional copies of this book please visit your local branch of any of the following world-wide chain bookstores:

- Barnes & Noble
- Chapters, Coles, Indigo

In the absence of shelf copies at your local bookstore, you may still order this book at any "Barnes & Noble" or any "Chapters – Coles" order desk via their in-store database "search & order" PCs.

This book is also available online at:
Amazon.com
www.iuniverse.com
www.bn.com "Barnes & Noble"
www.chapters.indigo.ca "Chapters – Coles – Indigo"

iUniverse books may be ordered through booksellers or by contacting:
iUniverse
1663 Liberty Drive.
Bloomington, IN 47403 USA
1-800-288-4677

About the author

Dr. Gouda's scholastic background presents a wide scope of professional scientific disciplines.
After first attending Medical school, he felt that becoming a GP wasn't his calling, and as he poetically put it, he just could not resist the laboratory appeal, the test-tube feel and his passion for analytical chemistry.

He switched to pharmaceutical research. His graduate studies at Masters and Ph.D. levels were in chemical toxicity and analytical medicinal chemistry.

Dr. Gouda is a renowned research chemist with several analytical SOPs named after him. The combination of his initial medical training, along with his extensive analytical chemist qualifications, has given him a unique ability to see beyond what a mere chemist or MD could see.

Dr. Gouda's experience includes:

- Medicinal research & development associated with the chemical complications of four specific pharmaceutical compounds.

- Pharmaceutical research associated with the drug treatments and prevention of four different health issues.

- Chief chemist of O.G.C. research Laboratories for over a decade. His experience also includes 6 years as a senior analytical chemist with Barringer research and EPL laboratories.

- Author of two university textbooks on analytical chemistry.

- Has written several papers on inorganic ultra-trace analysis via atomic absorption VS inductive coupled plasma.

- Accredited with the invention of 3 new patent-pending pharmaceutical compounds in the fields of hormone function manipulation.

Dr. Paul H. Ramses Gouda is a single parent to one child. The author and his son reside on Vancouver Island, British Columbia, Canada. He has recently partially retired, while still perusing business opportunities in the ownership of two enterprises, a hospitality business, and an analytical chemical commercial laboratory operation.

He is also a respected published author in a variety of other arenas.

www.goudabooks.com
www.optimumgreen.com

Front cover by **Pelé Gouda**

Graphic design, cartoon illustration and video animation.

pelegouda@gmail.com

Introduction by the author

Many people still express surprise and disbelief when they are told: "yes, you have a choice."

When I decided to end the "bachelor" and the "professional student" chapter and I decided to have the son I have always dreamed to have, I was not about to take a chance on not having a boy. Nothing against having a girl, absolutely nothing. It's just that I, personally, wanted a boy. I must admit, if I had a girl, she would have been a spoiled daddy's girl until she had reached the young lady stage of high school, when my protective orthodox personality may have just made her a miserable kid!

It was a rather late start on parenthood.
The mother – my ex-wife, had been married before, had three girls, and she had her "tubes" tied. The plan thus was one shot at one child, to be delivered via "cesarean" surgery at which time her tubes would be tied again for good. Hence, that one attempt, one child, had to be one boy.

In my immature, short-sighted macho attitude, I recall commenting: "what am I going to do with a girl, play Barbie?"

I wanted a boy to name him "Pelé" after the great footballer, and to get him into football ("soccer" in Canada & USA – "football" everywhere else in the world – and on this unrelated sports note, American football is "handball" everywhere else in the world.)

Well, I got my boy, I named him Pelé, he turned out to be the perfect son: great morals, graduated high school at 15 with honors and is probably the youngest student at his university … but, fate had the last laugh. He is no footballer. He is a video game geek. Well, you can't have everything.

I miss the days of feeding my baby, carrying him in the mall, the bedtime story, walking him to school - and I am now ready for

my next child, a girl this time. Her name will be "Fayrose."
I just have to find her mother first!

Yes I am just as confident that my next child will be a girl as I was confident my only child was going to be a boy.

That was 16 years ago. Yes, this is not a new discovery or a medical secret! In fact, many parts of the complete picture of this approach have been practiced throughout the world for centuries. Even the diet factor in determining sex selection was addressed in a 1930s book. Most of such information was practiced in Egypt, China, India, Eastern Europe and Ghana for literally thousands of years.

It is basic physiology, basic and simple facts, not someone's "method" – and, any medical doctor – not necessarily a gynecologist – would know all this.

Sex pre-selection in mammals in general can be guaranteed.

In fact, the same approach can be applied to animal breeding.
I first practiced it over 25 years ago at school, and for the past 10 years (including 5 years of artificial insemination) I have utilized it in breeding Great Danes, with an astonishing 100% success rate. Yes, 100%. I'll explain.

A few years ago, I announced on my Great Danes' site, 5 months prior to breeding (and I did the breeding via A.I. "Artificial Insemination") – that the next litter would be all males. I raised a few eyebrows of course, but as well as a litter of 11 pups, 11 males.

The following litter, over a year later, was again announced online that I'd be aiming for an all female-litter, as requested by applicants. And it was a litter of 10 puppies, 10 females.

Assisting a Chinese couple who wanted a boy, and who followed my instructions religiously, I was confident enough to place a congratulatory ad in the local paper, announcing their yet-to-be-conceived boy. About a year later, they named him after me.

And, I have the testimony of over 20 other couples I advised and assisted as requested. Some wanted a boy, others wanted a girl. Yes, 100% success, all of them, each and every one of them got the baby gender they wanted.

This is an important subject with major implications on the family's life and on society at large.

Ironically, I found out that this important subject has not been well addressed. A 1962 book, by an Egyptian doctor, a book by an American doctor in the 1980s and later a book by a British author, all simply put together the basic facts and collected data produced by others. They admitted how every point they made was known and utilized in the past, published by others, even decades earlier, and was supported by data produced by others also decades earlier. Yet, each one of them dared to claim the basic, public health information package and the scientific facts they reproduced as "his or her method."

Experimenting on an established facts and reconfirming it again for oneself, would not make it one's method.

The material of these books is basic, using generally correct facts that were paraphrased, quoted and represented, including charts that were cropped or redone, and complete sections that were simply hardly paraphrased. And that's fine, as long as they would acknowledge that they are presenting the public with general health facts and information, not a "method" someone invented.

Ironically, none of these books claimed more than 75% success rate! God knows what the real success rate is! Now, even 75% means that 1 out of 4 will fail. Not very impressive. One book bragged about a 66% success result by an independent Asian clinic. Surely is not impressive. I believe where two of the authors went wrong, is in two significant areas:

- They advised intercourse 24 hours prior to ovulation when trying for a boy. I disagree. This is too long a waiting period for the Y-sperms. It also eliminates the objective of offering sperm –Y a monopoly, as it allows excessive time for sperm –X to catch up, making it a 50-50 battle. I'll explain.

- They reported that artificial insemination has a high failure rate. I disagree. A.I., when done right, has a much greater success rate.

I had 100% success rate in both human conception cases as well as recorded animal breeding presented by my paper on A.I. in mammals. In fact, later on, it was also a complete success when the couples were only shown how to perform A.I. and they performed it in their own bedroom.

Recently I found a terrible book by a woman who is not a doctor, not a pharmaceutical chemist, nor a research scientist. I don't understand how she dared to address such an issue. It is full of contradictory and plain "wrong" information; including basic facts she didn't seem to understand.

I believe she misinterpreted a research paper done by a European doctor on a specific application. She clearly just collected some material from here and there, including, it appears, a piece of information taken from the back of an ovulation kit package that has been discontinued!

I couldn't help but feel terrible for all those innocent victim readers who will be misled and disappointed. This is just wrong! Besides, she did not claim a success rate above 60% which is only 10% over nature's odds of 50-50-%

Another book that was brought to my attention is basically a repeat reproduction of collected facts that were simply put together. It quotes experiments conducted in several countries. Yet again, it was also named someone's method!
As well, extracts from a couple of books on diet for sex pre-selection were brought to my attention. One is European, the other is American. The European one is quite old, while the

American book is fairly recent. They both presented basic facts that have been practiced in Africa and Asia for centuries, and they both presented specific diets, a one month and a two month diet.

One of them made the mistake of a one month diet to be repeated as is, i.e. day 1 & 31 would be the same and day 30 & 60 would be the same.
Both books made calculation errors in balancing the safe limits of what the body needs, and the required effective calcium, magnesium, potassium and sodium ratio. The last week of the diet should not be the same as the mid-week, i.e. day 30 & 60 should not be the same. The last week requires a different scale, and requires a specific corrective diet for the week following the pregnancy. Hence, I do not recommend either book.

I suggest there isn't a single factual, illustrated – and complete book on this subject, written for the average layman, and with sound personal experience and solid, justified potential success rate.

The 57%, 66% and 75% claims by these three books, are not good enough. I know what I am talking about. I'll show you how; with solid techniques, you can achieve a 90% success rate in some cases. I'll show you how, with additional specific applications you can actually secure a ≥99% success rate – very well, for all mathematical concerns, rounded to a perfect 100%.

And, I say 99%, not 100%, just to be theoretically sound, only considering the theoretical ridiculous, unreasonable argument that, say, a galaxy shift of energy imbalance could cause a system collapse tomorrow, placing several planets between us and the sun, and as of tomorrow, earth will become a dark planet of nights. Theoretically arguable. In reality, I claim a personal 100% success rate.

Having said that, the implementation of Conditioned A.I. technique I contributed in this field is certainly both unique and critical to increasing the success rate, beyond what is considered reasonable or good expectation.

Here is my contribution on this topic.

And here are the plain facts. They are not a method that I – or anyone for that matter – has invented; they are facts, as basic as any health information you should know. And, they are reflective of solid personal experience. Actually, if any one is going to have the bragging rights of calling it his method, I shall not be humble about it. It would be I. I didn't just present basic facts about the spermatozoa's characteristics: one is faster, one lives longer and accordingly, the elements of distance from the egg and the acidity of the vaginal mucus! There is more to what I'll present here. There is the diet factor and the employment of C.A.I. "Conditioned Artificial Insemination" technique.

I'll further comment that though the publisher has inserted a traditional copyright claim, in my opinion this refers to the book as a project in terms of production, not in terms of any single stated fact. No one has the copyright over the fact that "Y" spermatozoa is faster than the "X" one, any more than claiming copyright over the fact that the sun rises and sets. In fact, you're welcome to make use of any chart or diagram I personally designed.

What is different about this book is the illustrative presentations, the supportive personal experience, and the corrections made to other previously written books by others. As well, this book presents material that has not been covered elsewhere - such as the specific home utilization of diet and the conditioned artificial insemination – sperm separation - as an optional part of the complete information package.

My specific diet and supplementary formulas are another pioneer effort. Having said that, I must clarify that my sole drive here is not scientific community recognition – nor to ruffle anyone's feathers. The objective is to share my experience with the community at large.

I'll first present you with the basic scientific facts – some of which others have failed to see. I'll continue to introduce artificial insemination, diet and sperm separation, which will then bring you finally to the promised 100% success rate.

Every single piece of information presented in this book, is not just supported by the author's personal experience, they are also the scientific facts and research conclusions confirmed by at least 4 different sources over several decades.
They have been published by universities and scientists' papers in over seven countries. The "Conditioned Artificial Insemination" specific practice briefly indicated at the end of the book, reflects pioneer research by the author, currently being submitted to three universities, as a step towards hospitals endorsement.

The work is the result of a decade of mammal sperm treatment prior to A.I. and is presented solely as a scientific and a social contribution, made available to physicians and to the public with no remuneration expectations.

The detailed technical paper, including references – as well as the specifics of my data on both "Conditioned Artificial Insemination" techniques, and simplified effective "Sperm Separation" techniques, will be produced as a textbook following the official publication of my paper.

Finally, please also note that some inserted paragraphs were written in the third person, in reference to the author, as they were dictations of thoughts I wished to preserve. These notes were given to my assistant back then, and the publishing editor has deemed them, as presented, to be in favour of the flow of readability.

I hope I have provided the reader with a complete presentation, supported by both, scholastic training and personal practical hands-on experience.

Choosing the sex of your baby, or puppy?

Oh, absolutely. Indeed you can.

Dr. Paul Gouda

CAI & SSI

"Conditioned Artificial Insemination" &
"Sperm Separation Insemination"

This book started as a paper on A.I. "Artificial Insemination" in mammals, back when the author was a young medical student. It was later complemented by a follow-up paper by the author on two issues:

- The utilization of simplified, yet effective, semen treatment techniques prior to A.I. in order to favour one type of sperm over another. It was called "Conditioned Artificial Insemination" – or C.A.I.

- The utilization of simplified sperm separation techniques for the purpose of sex selection via A.I. The selected sample is inserted into the female (not a test tube fertilization.)

That was nearly three decades ago.

Not impressed by the prospect of leading a GP life, the author switched to chemistry; graduate analytical – pharmaceutical research, and led a chemist career for over two decades.

After consistent encouragement by friends who have been assisted by the author, the paper was then developed into the book you have purchased.

The book introduces to society at large, the concept of CAI "Conditioned Artificial Insemination" and SSI "Sperm Separation Insemination" - as part of the complete package. A part that can be simplified and implemented by ordinary couples – with the necessary medical supervision.

Any person interested in participating in a repeat documented - supportive study as part of a research paper on the effective utilization of Conditioned Artificial Insemination or Sperm Separation Insemination, may contact the author to volunteer his/her participation. Two clinics are planned:

The red tape technicalities, the academic facility arrangement, and the details & terms of participation are being studied and will be presented at that point.

The planned study is meant to provide an official "confirmative paper" – solely as part of the scientific routine.

We intend to repeat this confirmative exercise in three cities (yet to be selected based on residence of volunteer applicants.)

We have an absolute confidence that the independent results will confirm and support the incredible success rate we have achieved thus far.

One more comment: I attempted to satisfy the scientific logic or support for the technical aspect of the presentation, as well as offering a book for the layman. I felt, while presenting simple material, I should also present, as a bonus, reasonable supportive technical notes. Personally, when I read a book, I expect the author to defend and support his hypotheses. I expect him or her to try to convince me of each hypothesis' claim and to show me its own foundation of scientific logic or to give at least a simple explanation. I like to see "how" and "why" and to understand the logical justification behind it.

Well, I was advised to change course. Three proof-readers from three walks of life felt that the technical explanations were too technical. I listened. I deleted over 70 pages of scientific explanatory notes, as well as 30 other pages of artificial insemination material. Very few technical notes were left in this book. They are, I felt, necessary, especially when I addressed the diet factor and for example the role of free radicals; and when I addressed the chemical manipulation concept of the OGL products. Having said that, even these few technical notes here and there can be skipped by the reader. Otherwise, the material was kept simple and straight forward.

I trust you'll find this book a helpful reference.

To my only begotten son, Pelé - who, more than once, has assisted me with performing “Conditioned Artificial Insemination” on mammals.

Table of content

Yes you can choose,
and with certainty.

My own personal mammal record,
both human conception and animal breeding,
presents 100% success.

Part – I

The basics, presented correctly, clearly and illustratively. And more.

The conception story…

… The tortoise and the hare.

In a rather poetic opening, let me tell a story about miss Ovum (known as the egg) – who walks into a well prepared bedroom, lies on a silk bed, surrounded by servants, she awaits passively, for the aggressive sperm that will rudely attack her.

And Mr. Charming, the sperm who knows nothing of charm, has to go through many obstacles. He has to climb mountains, swim across an ocean, fight barbaric foes, and survive either a cold winter, or a hot summer (depending on the condition of the host as I'll explain) – to finally penetrate Miss ovum without even saying "hello" first.

There are different kinds of sperm, not just two as you would read in some books. However when it comes to impregnating that ovum, there is the "Y" chromosome sperm, which will create a boy - and the "X" chromosome sperm, which will create a girl. The ovum, or "egg" will accept whichever gets to her first, and then she locks the bedroom door allowing no one else into her. She will however try to give a helping hand to one sperm over the other if she was fed her favourite food.

So, it is the man's sperm that decides the sex of the baby, not the woman's egg.

The question now is: How to control or assist either the "Y" or the "X" sperm to get there first, or, how to prevent one of them from getting there.

We said there are different types of sperms. Sperm whose job is a bodyguard protecting the strong "Alfa" sperm, and fighting any other sperms found that were donated by a different man. Sperm whose Y- X characteristics are not defined, sperm with abnormal characteristics - and of course, we have the active boy-producing Y-sperm and the active girl-producing X-sperm.

We'll address two approaches in this manual. One that would increase the chances – greatly – in favour of either Y or X sperm, and the other approach will eliminate either Y or X entirely. I'll explain why sometimes one approach is more practical than the other.

Well, let's first look at the basic concept of the first approach:

The distance factor & intercourse position:

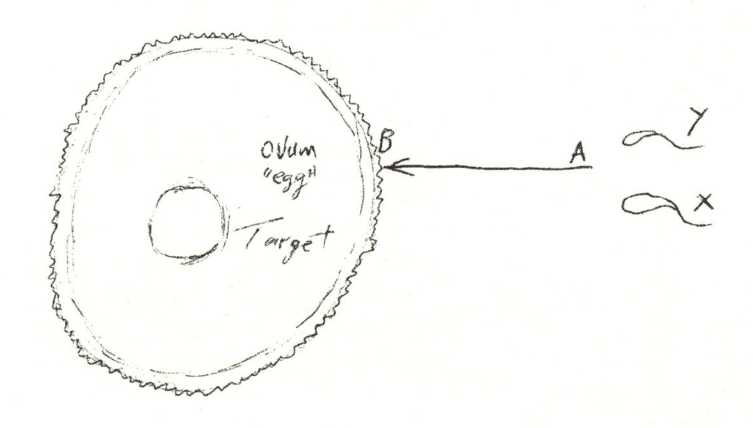

The distance from "A" to "B" is the journey the sperm, both "X" and "Y" will take from ejaculation or deposit point to the egg.

Over 100 million sperm will start at point A. Only about 100 or less will have a chance to fertilize the egg. Less than one in a million will make it. This fact emphasizes the nature and the difficulties of the environment, or the journey, from A to B.

Whichever sperm that makes it there first, X or Y, will penetrate the egg, creating the embryo "fertilized egg," and no other sperm will be allowed in.
The whole approach of how to significantly increase the chances of a specific type of sperm to reach the egg first depends on these simple facts:

1- The Y sperm, boy-producing sperm, is slimmer, faster, but with less staying power and as such, a shorter life span.

2- The X sperm, girl-producing sperm, is fatter, slowr, with greater staying power and better longevity.

3- The chemical composition at point "B" around the egg surface, will favour one sperm over another. This can be controlled by a dietary and supplementary programme and can be eliminated as a factor by sperm elimination or sperm separation techniques.

We can call the Y sperm a skinny short distance sprinter. He will be ahead, faster within a short distance, then will run out of breath, slows down, eventually stops and falls down on the shoulder of the road, dead.

While the X sperm is a choppy marathon runner that takes it slow, steady and easy. It runs slower for a longer distance and will last longer. It will be behind in the first stage, but will continue, slow & steady, for a longer distance, and for a longer period of time.

Well, this means that, the shorter the distance, the better the Y-sperm chances to make it first to the egg. Remember the fable of the tortoise and the hare!

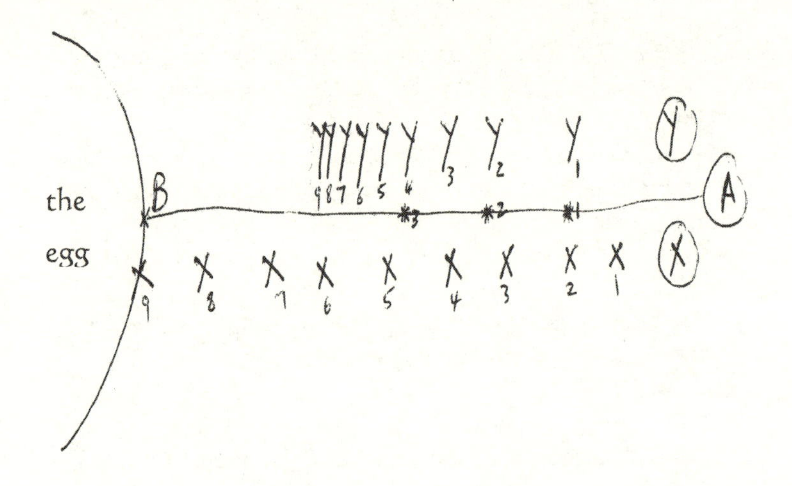

Let’s say we have 2 sperms, one Y and one X, both are starting at point “A” and the target is point “B” where the egg is waiting.

The distance from A to B is a race between the fast & slimmer “Y” vs the fatter, slower “X”.
Both sperms are illustrated above, over specific and equal periods of time labeled as 1 to 9. Meaning, points 1 to 9 show where the same sperm would be at point 1, 2, 3 to 9 at any point in time.

At the moment when Y would be in position Y1, X would be in X1 and when Y is in point Y5, X would be in X5.

If you look at the “AB” distance, you find 3 sample checkpoints marked for illustration: *1, *2 & *3

So, both Y & X started, of course at the same time, at point A. On their way from A to B, at point *1, you see that at that moment Y1 was ahead of X1. That means if the egg was located in point *1, sperm Y would have been there first, and sperm-X would be behind in position X1.

The race line from A to B continues. In *2 (on the AB line) you see that Y2 is ahead of X2. At point *3 you see that Y4 is ahead of X4 by a shorter distance.

As time and distance increase, the faster, slimmer sperm-Y slows down until he runs out of breath, and the slower, steady sperm-X catches up to him, and slowly passes him leaving him behind – as we see in position X6 Y6.

This gives us the first elements of the complete picture:
Distance between the ejaculation point and the target point.

The shorter it is, the better sperm-Y's chance to be there first producing a boy.

This translates to: **intercourse position**.

In order to shorten the AB distance and to give the Y-sperm an advantage, deep penetration and deep deposit are needed.

Place a pillow under the woman's "gluteus maximus" or butt cheeks, and no pillow under her head to create a flowing slope angle in favour of the ejaculated semen movement, and place her legs up on your shoulders whereby the back of the knee areas would be resting on your shoulders. Deep penetration and a deep deposit point is the objective.

On the other hand, for a longer AB distance giving X-sperm the advantage, aim for a shallow deposit avoiding deep penetration, i.e. farther away from the cervix, right into the acidic vaginal secretions. From a missionary position, place a pillow on the lower area of her tummy, above the genital area.

I believe there is no need here for cartoon illustrations here!

Notes:

Synopsis

- Y-Sperm is slim, fast, less hardy, short-lived (boy producing sperm)
- X-sperm is fat, slow, lives longer (girl producing sperm)
- Shorter distance favours the fast Y-sperm.
- Longer distance favours the hardy X-sperm.

Condition of the vaginal environment.

The next factor is the condition during the AB journey of the sperm.

We said that sperm-Y, the boy-producing, is fast, slim, with less stamina. While the X-sperm, the girl producing sperm is slow, chubby with more stamina (potential to live longer and to survive a harsher environment.)

These are facts known to us, and they have been mentioned by dozens of magazines and papers all over the world for decades.

We also know that the woman's vaginal environment is generally acidic. This makes it difficult for the Y-boy sperm to survive for a longer period of time. The fatter, slow X-girl sperm will live longer.

However, on ovulation day, the woman's environment (cervical mucous) is alkaline, which favours the weaker, faster Y-sperm.

This means that the timing of coitus in relation to ovulation is significant.

The sperm deposited prior to ovulation can survive for a few days waiting for the egg. However, if the sperms are deposited 2 or 3 days prior to ovulation, they will have to sit around waiting for the release of the egg (ovulation), in a harsh environment, which means that, the weaker-faster Y-boy sperm will die first. Then there comes the egg and it's a sitting duck for the hardier girl-producing X sperm that were able to survive the acidity – and other elements for a few days.

Meanwhile if the sperm is deposited just before ovulation or just as the woman ovulates, the environment is alkaline, which makes it easier on the Y-boy sperm to utilize the advantage of its speed - again, subject to the previously explained "distance" factor.

We'll talk later about how to determine the ovulation date.

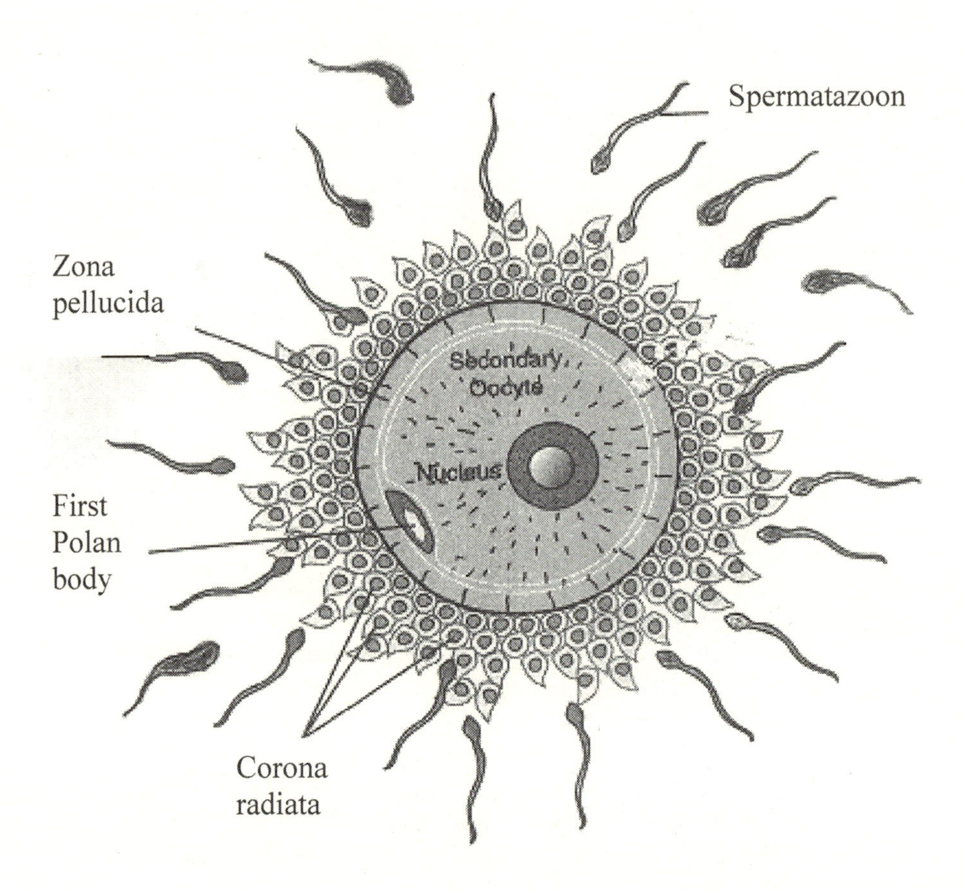

I'll also talk later about the diet & supplements factor in influencing the sperms and the egg at the point of contact.

The factor that must be emphasized at this point is, for a boy, intercourse should be just prior to, or at the time of ovulation.

For a girl, it should be two to three days prior to ovulation.

I must again mention that this factor alone will contribute an average of approximately 10% advantage. But, that is a considerable advantage. To significantly increase the odds you would need to implement ALL the elements, or parts of the puzzle.

And, I wish to comment here about the use of douches. I used to make my own low pH (acidic) douches with basically a mixture of any of the following: lactic acid, carbonated water, filtered lemon juice and vinegar in order to increase acidity.

I used baking soda douches (high pH = low acidity) to increase alkalinity - when I bred Great Danes. 3% sodium bicarbonate solution, followed by a 1% of sodium bicarbonate. This will favour a boy conception.

A simple 30g baking soda in a liter of water for a boy, or 30 ml of white vinegar in a liter of water for a girl.

This simple wash just before intercourse may help, but this becomes less significant with artificial insemination when trying for a boy - as a deep deposit starting at the less acidic cervix area will bypass the acidic vaginal trip.

Generally, for human conception, the idea is to use an acidic douche (low pH) when seeking a girl, and an alkaline (high pH) douche, to increase alkalinity or decrease acidity, when seeking a boy, only if necessary. Do it only if the woman's acidity, for example, is abnormally high.

The acidic environment is the default condition in general, and any increase (e.g. use of vinegar) could negatively influence all sperms. Meanwhile decreasing acidity can be adequately achieved with a repeated mere water rinse (pH 7). In fact, the use of ovulation date determination is a much more significant factor.

As indicated, ejaculated sperm deposited deep into the back of the vagina will have a better chance to make it to the egg. Secretions near the cervix (the opening into the womb) are generally less acidic, if not actually alkaline.

Hence, aiming for a boy, a deep deposit close to the cervix creates a combination of two advantages: a shorter distance to the egg, and a more alkaline environment. It bypasses the first obstacle of vaginal secretions where most sperms die.

As I'll explain later, artificial insemination (at home) is the final part of the complete picture. It'll maximize the odds to achieve near that $\geq$ 99% success rate I personally achieved for over 10 years with all who approached me- as well as at home, in animal breeding. And yes, A.I. principle is the same very well in any mammal, humans are no exception.

I do acknowledge of course the somehow smaller sample data. However, based on a careful mathematical calculation, taking into account the probabilities of failure in each factor – and provided that all factors are properly addressed, we still achieved a good graph correlation sufficiency of over 0.98.
It is my finding that a success presentation chart would still be higher than 90% - well above any previously reported average.

I mentioned elsewhere that I had announced on my Danes site that the next litter would be all males. They were. The following litter, I announced that, as requested by applicants, the puppies would be all females. They were.

I planned to have one child, a boy. I did. I assisted dozens of friends to have the child they wanted, and not one single case was a failure. And I didn't do it for money. I didn't charge any of them a single piaster, nor was I practicing medicine at the time as a profession; I was a chemist.

Back to our presentation … the following textbook drawing will help you visualize the trip to the egg.

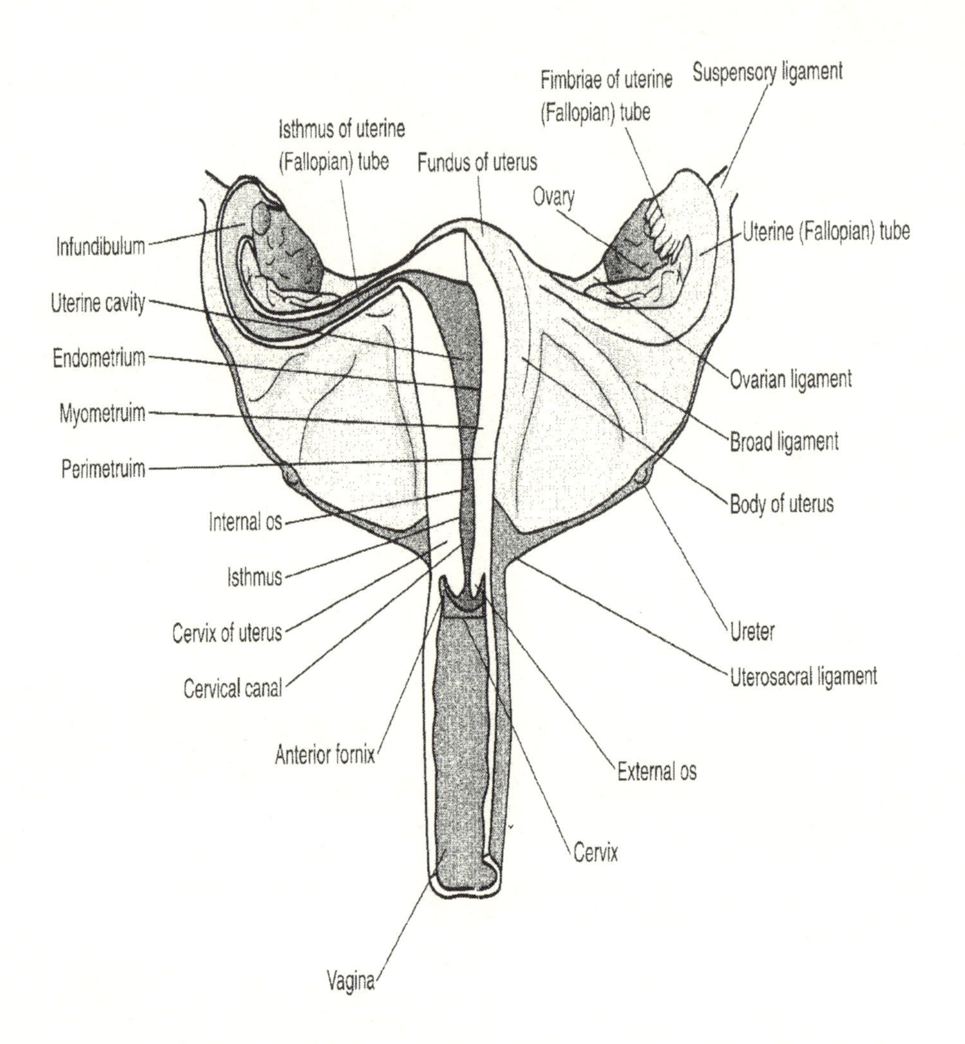
Fimbriae of uterine
(Fallopian) tube
Suspensory ligament
Isthmus of uterine
(Fallopian) tube
Fundus of uterus
Ovary
Infundibulum
Uterine (Fallopian) tube
Uterine cavity
Endometrium
Ovarian ligament
Myometruim
Broad ligament
Perimetruim
Body of uterus
Internal os
Isthmus
Cervix of uterus
Ureter
Cervical canal
Uterosacral ligament
Anterior fornix
External os
Cervix
Vagina

Synopsis:

- Vaginal secretions are acidic by default. They become more alkaline upon ovulation.

- The acidic environment favours the hardy X-sperm.

- At the end of the vagina, near the cervix, the pH is higher, i.e. the environment is more alkaline – or less acidic.

- For a boy: time the intercourse very close to ovulation.

- For a girl: 2+ days before ovulation.

- For a boy: use a deep penetration position. This will shorten the distance to the egg, which favours the faster Y-sperm, and will bypass the harsh acidic vaginal secretion trip.

- For a girl: use a shallow penetration position. This will help eliminate the less hardy Y-sperm that will die in the acidic vaginal secretions. Also, longer distance will give the slower X-sperm a chance to catch up.

Who gets the power in numbers advantage?

This brings us up to the next factor: **The numbers game.**

The larger the starting number of sperms the more advantageous it becomes in favour of the faster Y-sperm.

This would be like taking one man to “box” or “kick-fight” one woman. On average the man is physically stronger, but you could have the exceptional weaker man and the exceptional stronger woman. However if you widen the scope, a million men against a million women, the one final single champion will be a man. Physiologically, nature has assigned physical strength to the man to do the hunting, fight the beast, come back to the cave with the kill, to feed the woman and offspring. This genetic muscular disposition has not been changed despite modern social concepts.

In other words, if you were to change the world boxing championship to a one unisex competition, certainly no woman will ever win it. Very few women, if any, would come close to challenge say, Mike Tyson, yet thousands of men would.

Similarly an individual female sprinter might, in an individual case, be faster than an individual athletic male, but a thousand male sprinters against a thousand female sprinters will produce a male in first place.

The same principle applies here.
If the quantity of the sperm deposit, or ejaculation is great, the chance of one of the faster Y-sperm making it to the egg is greater.

If the starting number of total sperm is relatively low, especially over a longer journey, the success chance of the more resilient X-sperm is greater.

The advice to men trying for a boy, would be:

- refrain from sex for a week
- excite yourself and psych yourself up, perhaps fancy being with your most beautiful superstar at the moment of ejaculation.
- achieve deep penetration (I mentioned the preferred sex position.)
- allow maximized ejaculation.

We'll talk about the diet factor later.

And if seeking a girl, the opposite … for example, as soon as ejaculation starts, picture your wife being the ugliest most repulsive woman you know, think of football or politics, and pull out slowly.

And of course, when we talk about a smaller ejaculation, we still mean within the acceptable or the normal range; not to the point of an abnormal low sperm count. The reason is, multiple sperm are required to break through the zona pellucida, before that one successful sperm can penetrate to fertilize the secondary oocyte.

The goal is to secure a quantity of Y or X sperm to get there first. A sperm count under 25 million per cc has shown to produce mainly girls, while a sperm count of less than 3 million per cc will only produce girls.

Hence, to increase the chance of a boy, the man would need to adhere to a rich diet and to abstain or refrain from sex for about a week prior to the pregnancy attempt.

On the other hand, for a girl, the man should masturbate or practice safe sex the day prior to the attempted pregnancy.

We'll talk about specific diets later.

And again, this one factor alone is not enough. You would need to address all areas for the complete picture.

Synopsis

- For a boy, refrain from sex for a week. You need a deposit with a large quantity of sperm.

- For a girl, masturbate or use a condom the night before the intended pregnancy. A lower count deposit is favoured.

- Deep penetration position for a boy.

- Mental stimulation and excitement at the time of ejaculation for a stronger deposit when trying for a boy.

- Vaginal conditioning, washing with water or diluted baking soda, to reduce acidity when trying for a boy.

- Vaginal acidity enhancement, using vinegar to increase acidity when trying for a girl - "see notes presented by the preceding page."

Notes:

Woman's psychological & physiological input factors.

Well, since we are attempting to use every major and every secondary factor to increase the odds of success, we must utilize every proven element that may give nature a hand.

One of these secondary factors is the woman's mood.

When the woman is enjoying sex, foreplay was done right, and chemistry is passionate between the two of them, the woman releases secretions, referred to as getting "wet" - which helps not only reduce her acidity, but also influences the chemical medium surrounding the ovum in favour of a boy.

This fact reaches its optimum effectiveness if the woman achieves what is known as "female orgasm." You'll read in some books that the reason we want to achieve that is because at that point her acidity is minimal, and the vaginal environment will be more accommodating of the weaker but faster Y-sperm.

While this is true, the more significant factor is an old brain programme we inherited from the days of the jungle. We'll talk about these programmes later. Woman's excitement at that point is interpreted by the brain as a proof of happiness and prosperity. Or in other words, she has no worries, there is plenty of food, a safe cave and a strong protective man.

The brain reaction in this case results in secretions favourable of faster removal of a layer on the head of Y-sperm. Both Y & X sperms have a layer that must be dissolved before they can penetrate the egg's zona pellucida. The chemistry around the egg surface will favour one sperm over another as we'll talk under diet.

A prosperity signal to the brain favours male production to protect the prosperous territory. While lack of happiness or signs of physical famine would signal hard times, and race survival would require more females. Statistics show that most women on poor diets give birth to more girls.

The point is, the woman's psychological factor of welcoming or resisting the pregnancy, plays a role in gender selection.

I guess, and I am not really being entirely facetious, for a girl, one way or another, she should not get herself "in the mood" – perhaps you could remind her to think of an irritating thing you did, and keep it a "wham bam, thank you ma'am" performance.

For a boy, it should be a passionate love-making session, and it would be ideal if she experienced orgasm just before the man has ejaculated.

And, we'll talk in the next chapter about how the woman's physiology is influenced by diet and how diet influences the secretions in which the sperm performs the vital ovum penetration process. For a thousand years in Africa, they fed women boiled oysters, unsalted, to help her have a girl, and salted fish to help her have a boy. It wasn't until late 1800s and early 1900s when science realized the validity to these old tales.

This can be easily explained when we consider how high levels of sodium ions in the body excite the receptors attached to the cell membrane.

The activity of these receptors influences the ionic concentrations of the secretions, including those in between the head of the sperm and the egg. In the excited state, the receptors will attract available corresponding ions of the highest concentration of potassium, calcium and magnesium.

Woman's psychological factor will influence her physiology, and diet will give the specific sperm a helping hand.

Synopsis:

Female orgasm (ideally just before male ejaculation) - and her sexual excitement releasing vaginal secretions, results in reducing her acidity. It also results in a chemical development around the egg. Both changes favour the Y-sperm.

Nothing new about that, we have research that dates back to nearly a century ago addressing this issue.

Woman's psychological factors will influence her physiology. And woman's physiology related to diet will give either the X or the Y sperm a helping hand.

And again, remember the important elements of sperm quantity and the distance from the ovum. For a boy, aim for a deep and a large deposit. For a girl, a shallow penetration and a small ejaculate.

Notes:

Whenever additional technical notes were entered, you have the option to skip such paragraph or page. The simple instructions re-emphasized by the chapter's synopsis are very well all you need.
The technical notes were kept to minimum (over 70 pages were deleted from this addition. And, over 25 pages of illustrations and simple explanations were added.)

I encourage you however, to read this book more than once, including a third reading in its entirety.

The delicate nature of the Y-sperm

Would playing video games influence the man's sexual reproductive performance?

Would the clothes the man wears have anything to do with whether he'll have a boy or a girl? Come on now!

Well …
Remember that the Y-bearing sperm is delicate. Y-sperm is smaller with less staying power. They die easier and sooner than the more hardy X-bearing sperm (perhaps like later in life. Statistics show that women live longer than men!)

The sperm in the testes are under environmental pressure. The X- sperm will handle that pressure better than the Y- sperm can.

Relatively excessive heat will kill many Y-sperm before ejaculation – or will ejaculate them in a weakened condition that would make their chances of surviving the vaginal environment even slimmer.

Statistics prove that most boys are born during the summer (meaning that they were conceived during the winter when the testicles were not too warm for too long.)

So for a boy, wear loose shorts for a week prior to intercourse. Keep the testicles "aired" – perhaps even avoid underwear and avoid silk or polyester fabrics. Stay in air-conditioned environment until the mission is accomplished, so to speak.

And, those whose profession or excessive hobby exposes them to long hours of screen radiation, e.g. computer monitor or TV, again statistics or data published by Asian and South American scientific sources concluded that those video geeks, whether by choice or by profession, are more likely to produce girls. In fact

men whose job involves long hours of sitting down, such as a long distance truck driver, they too produce more girls.
For generations it has been said that a hot steam shower before sex will either make the man produce a girl or will act as contraceptive. There is validity to this statement.

Perhaps a reminder of what is expected from that tiny delicate creature would be helpful. The reason the man produces a ridiculous quantity of spermatozoa, is the very fact that very few will survive. While most men ejaculate 100 to 200 million sperm, a healthy man could ejaculate up to a billion spermatozoa. Only one sperm is needed. And only about a dozen or two will make it to the egg.

Perhaps a flaw in design. If I had been consulted with at the time of creation, I would have advised differently!

Back to our subject - the sperm will not receive the hospitable service and the convivial reception the ovum receives. The egg is surrounded by servants of cells serving her 5-star meals as she makes her way from the ovary into the fallopian tube. On day 14 of the menstrual cycle (the period,) ovulation occurs as the egg erupts from the follicle on the ovary surface, surrounded by thousands of nourishing and protective cells, carried on a convertible Mercedes (hair-like projections of the fallopian tube) from the tube into the womb.

The hormonal service did not forget to prepare a comfortable bed, a lining in the womb in anticipation for the arrival of the fertilized egg, and to provide her with early nourishment - (these are the cells that are expelled via the menstruation discharge if the egg is not fertilized.)

The large egg starts to remove most of its external "corona radiate" in what has been described nearly 80 years ago as the egg's act of undressing in preparation for the sexual encounter.

She doesn't have to do a thing. She'll lay there in comfort waiting for the winning sperm to penetrate her.

Now the tiny sperm does not have it that easy. It is thrown into a hostile acidic vaginal secretion where most of them die. The sperm struggles, using a tail to swim as fast as it could, towards the less acidic area near the cervix. It then has to overcome the current of sticky swirl of the cervix fluid. Then has to struggle towards the fallopian tube, responding to the egg's song of love, a molecular emission the egg sends.

Studies show that nature has compensated the Y-sperm that was made less resistant to acidity, with a better reception allowing it to make a better choice in selecting the right tube more often than the X-bearing sperm. Many more X-sperm enter the wrong fallopian tube and die there. Though at this stage it is not over, and many X and Y sperm will start the struggle in the right fallopian tube towards the egg, still struggling against the muscular contractions of the tube walls, the wall's hair-like cells that traps many sperms, and the downward stream of cells that assists to propel the ovum down.

Then, the female's white blood cells will treat the sperm as infectious invaders. No exaggeration when, in a perhaps a poetic tone, I said that the sperm would have to cross oceans, fight sharks, cross desserts with intolerable heat "acidity" and climb mountains against wild beasts before they finally make it to the comfortable egg, laying on a shady bed and surrounded by fruit trees. Well, close! The point is, the poor sperm can use a hand. Don't torture it before it has even been ejaculated. That's just bad parenting.

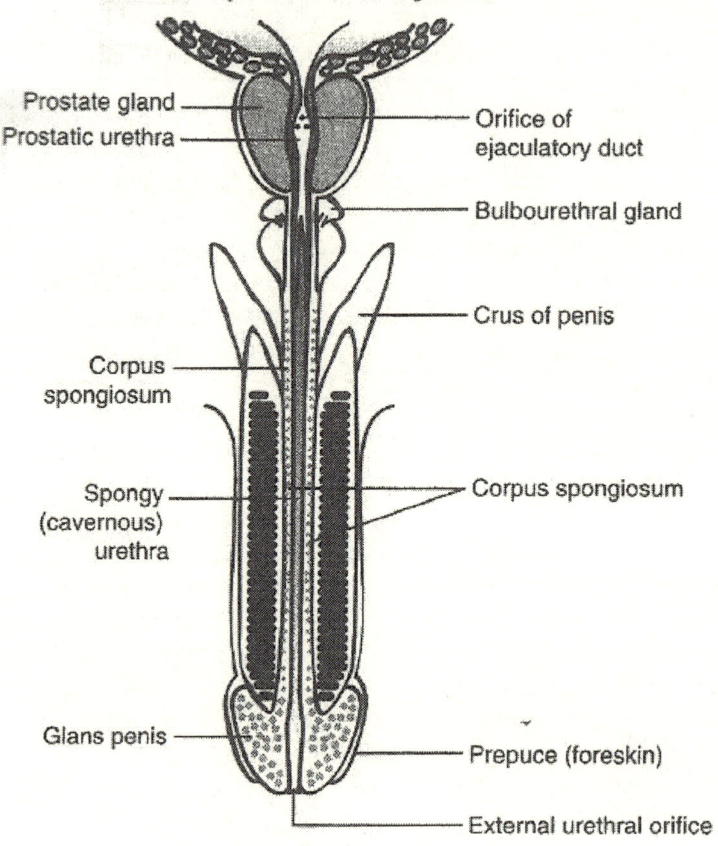
Male Reproductive System
Prostate gland
Prostatic urethra
Orifice of ejaculatory duct
Bulbourethral gland
Crus of penis
Corpus spongiosum
Spongy (cavernous) urethra
Corpus spongiosum
Glans penis
Prepuce (foreskin)
External urethral orifice

Synopsis

The delicate nature of the Y-sperm requires more attention in order to attend to their needs.

Temperature is a vital factor. If the testes are too warm, e.g. tight underwear, especially polyester material, which produces a harmful charge, it will reduce sperm count, especially the boy-producing Y-sperm.

The same applies to environmental hazards such as radiation from TV and computer monitors. Even your cell phone; don't place it in your trousers' pocket near the testicles. Avoid any radiation source as much as possible. In fact a new study on rats is the subject of a possible male contraceptive approach. It utilized exposure of the testicles to high frequency ultrasound which paralyzes or kills the existing sperm.

Food such as steak, fish and nuts are good for sperm health and production. Do not replace fresh fruits & veggies with convenient canned food. Canned food – as I addressed in the separate technical academic paper – is responsible for poor sperm health.

Notes:

The moment of truth

When the sperm reaches the egg in the upper part of the fallopian tube he attacks head-on, no foreplay, no dinner and a movie. He has suffered enough and having made it thus far, he has only one thing on his mind, the egg's inner sanctum.

Pronucleus is a cell nucleus with a haploid set of 23 "human" chromosomes resulting from meiosis (germ-cell division). The male pronucleus is the sperm nucleus after it has entered the ovum at fertilization but before fusion with the female pronucleus. Similarly, the female pronucleus is the nucleus of the ovum before fusion with the male pronucleus.

Again, the man always decides the sex.
The man provides a "Y" sperm or an "X" sperm.
The woman's egg is always and only an "X" chromosome.

The sperm Y chromosome "a boy," is the predominant chromosome.

The sperm X chromosome "a girl," is the recessive chromosome.

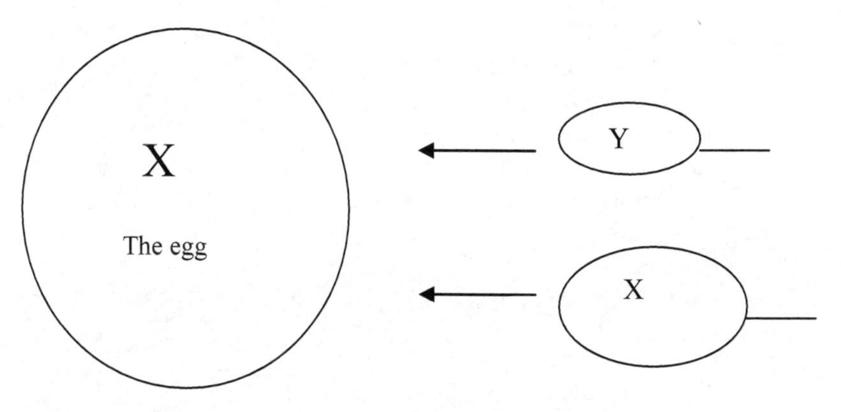

So, the egg is "X" …

if it was penetrated by a Y = XY = boy
if it was penetrated by an X = XX = girl

And on the subject of dominant genes, in animal breeding, here is a similar illustration, the issue of colour selection when breeding Great Danes:

The Brindle "B"colour is predominant, the fawn "F" is recessive.

The pup will take one colour gene one from each parent.

If the mother is fawn, then she is FF (fawn-fawn), carrying no B "brindle" gene.

If she had a B gene, the B gene would have dominated and "B" gene would appear, making her a brindle colour carrying BF.

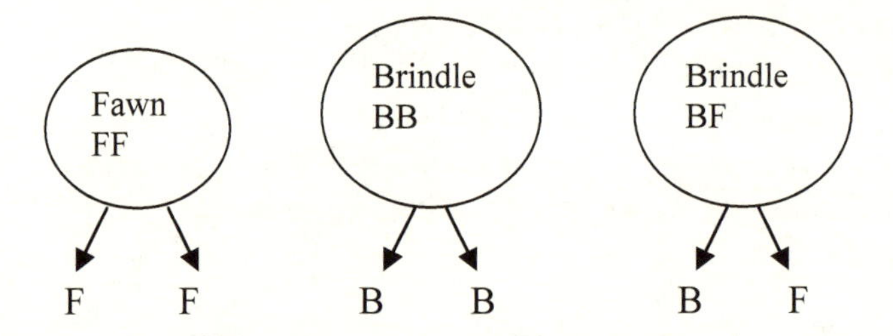

If the father is Brindle, he is either BF or BB.

If he is BF (brindle fawn) – and the mother is FF, there is a chance to have an FF puppy = fawn pup (meaning the pup took an F from each parent). Otherwise any BB or BF pup will be brindle in appearance.

If one parent is BB, even if the other is FF, all pups will be brindle because BF = B

That means if the Dane is fawn, he is FF for sure.
But, a brindle Dane could be BF or BB.

And that's why, since the B gene is the dominant one, if the brindle is BB and that Dane is bred to a fawn, which means bred to FF, then all pups will be BF, and they all will be brindle, but they will be able to donate F genes later when they are bred.

Notes:

The male and the female contribute 22 chromosomes each, to form all characteristics: eyes, hair … etc. except the sex. One separate chromosome from the male and one from the female will decide the sex.
A total 46 chromosomes.

The female: XX
The male: YX

The female pronucleus is the haploid nucleus of the fully mature ovum which loses its nuclear envelope and liberates its chromosomes to meet the synapsis with those from the male pronucleus.

The male pronucleus is the nuclear material of the head of a spermatozoon, after it has penetrated the ovum and acquired a pronuclear membrane.

Notes:

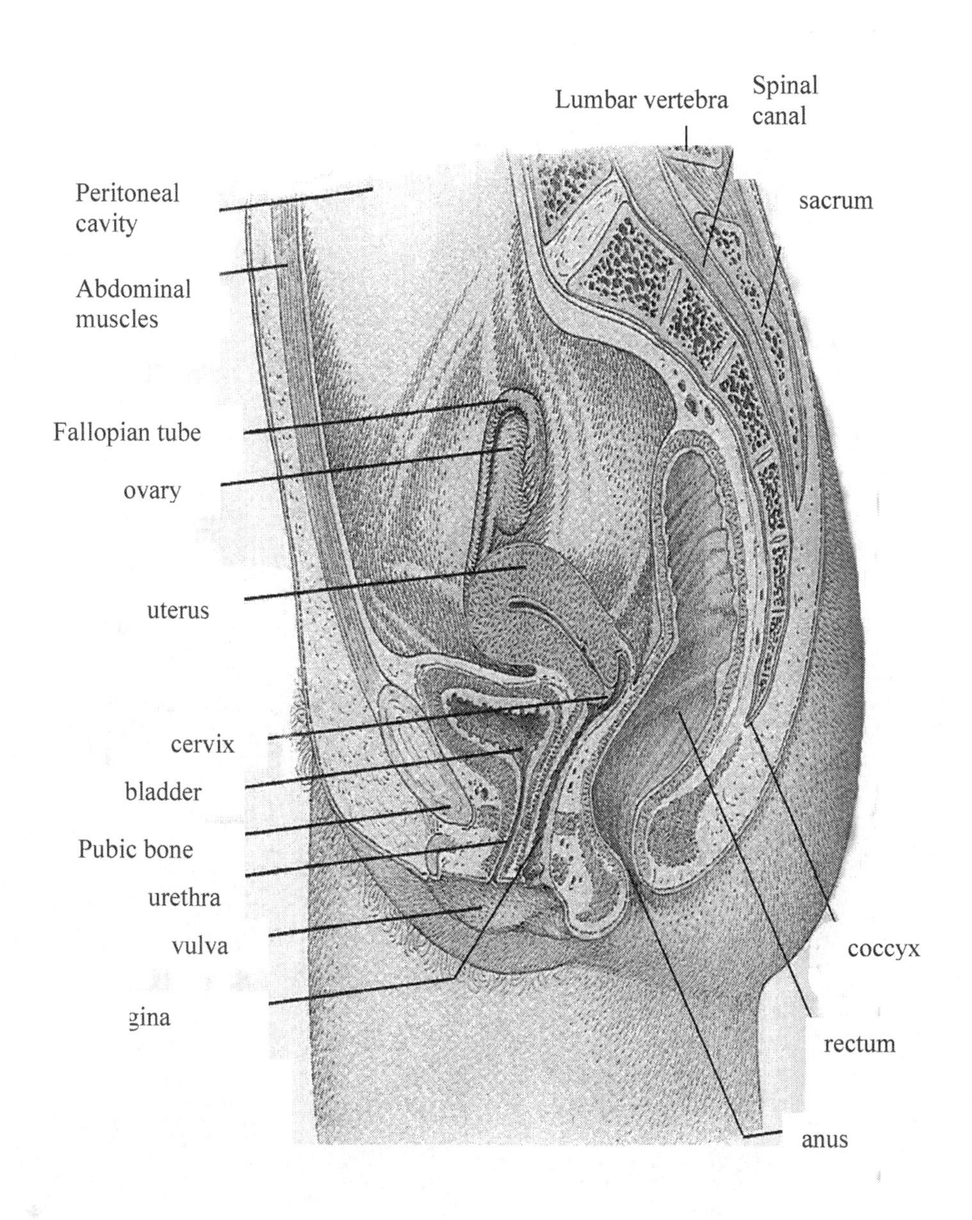
Lumbar vertebra
Spinal canal
Peritoneal cavity
sacrum
Abdominal muscles
Fallopian tube
ovary
uterus
cervix
bladder
Pubic bone
urethra
vulva
gina
coccyx
rectum
anus

Synopsis

- Boy sperm “Y” chromosome is dominant.

- Girl sperm X chromosome is recessive.

- The egg is ‘XX."
  The sperm is “YX.”

The new creation “baby-to-be” takes one sex gene from each parent, the egg, is always “X”, and the sperm that may carry either a boy “Y” or a girl “X” gene.

- For a girl, both donated genes must be X = XX = girl.

- One Y “boy” + one X “girl” = XY = boy

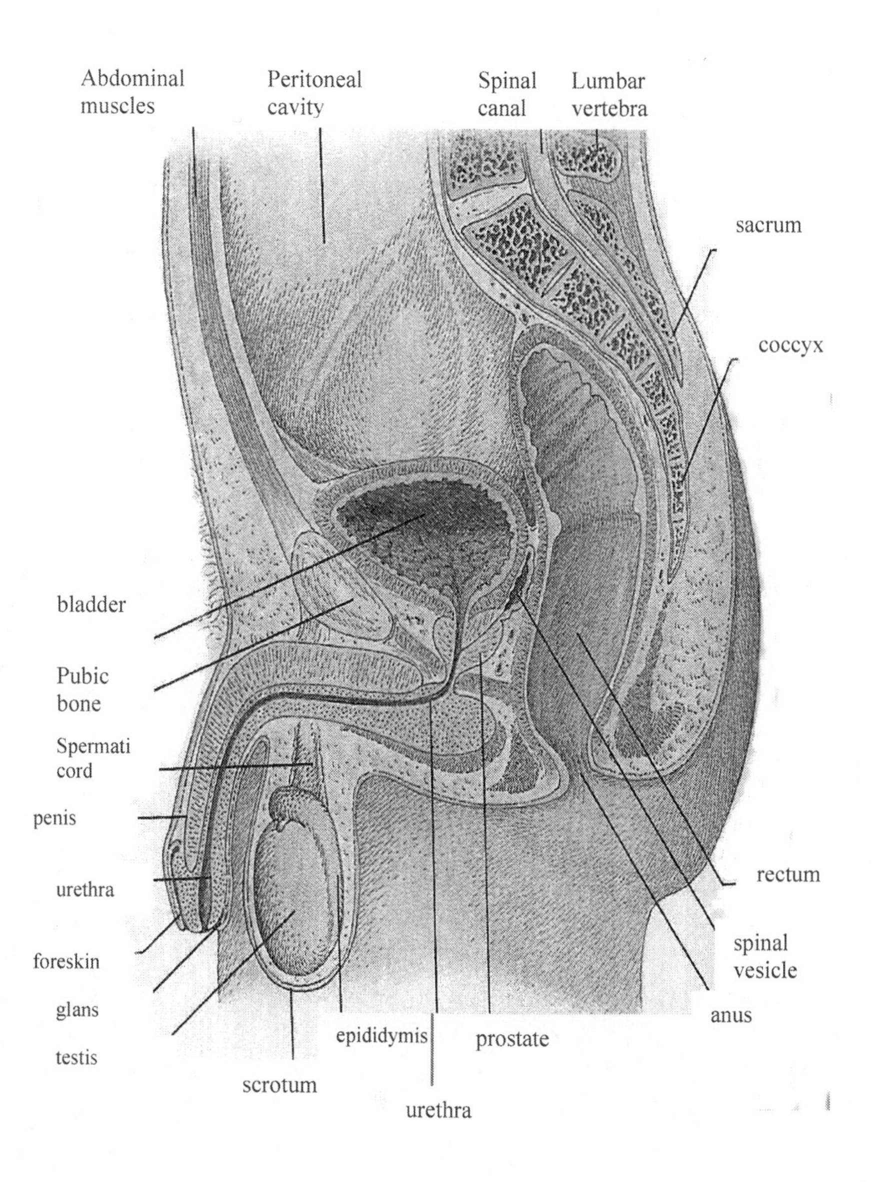
Abdominal muscles
Peritoneal cavity
Spinal canal
Lumbar vertebra
sacrum
coccyx
bladder
Pubic bone
Spermati cord
penis
urethra
foreskin
glans
testis
scrotum
epididymis
urethra
prostate
anus
spinal vesicle
rectum

Accurate determination of ovulation date is a significant factor.

It is an effective tool for the manipulation process of sex selection.

Ovulation date

Starting with the first day of bleeding, it's generally the 14th day. That makes it the middle of the average 28-day cycle. But, not every woman has a theoretically regular ovulation date.
We need to determine that date with accuracy for sex selection. The following medical textbook chart presents a common scenario – keeping in mind that a day or two here or there could differ from one woman to another.

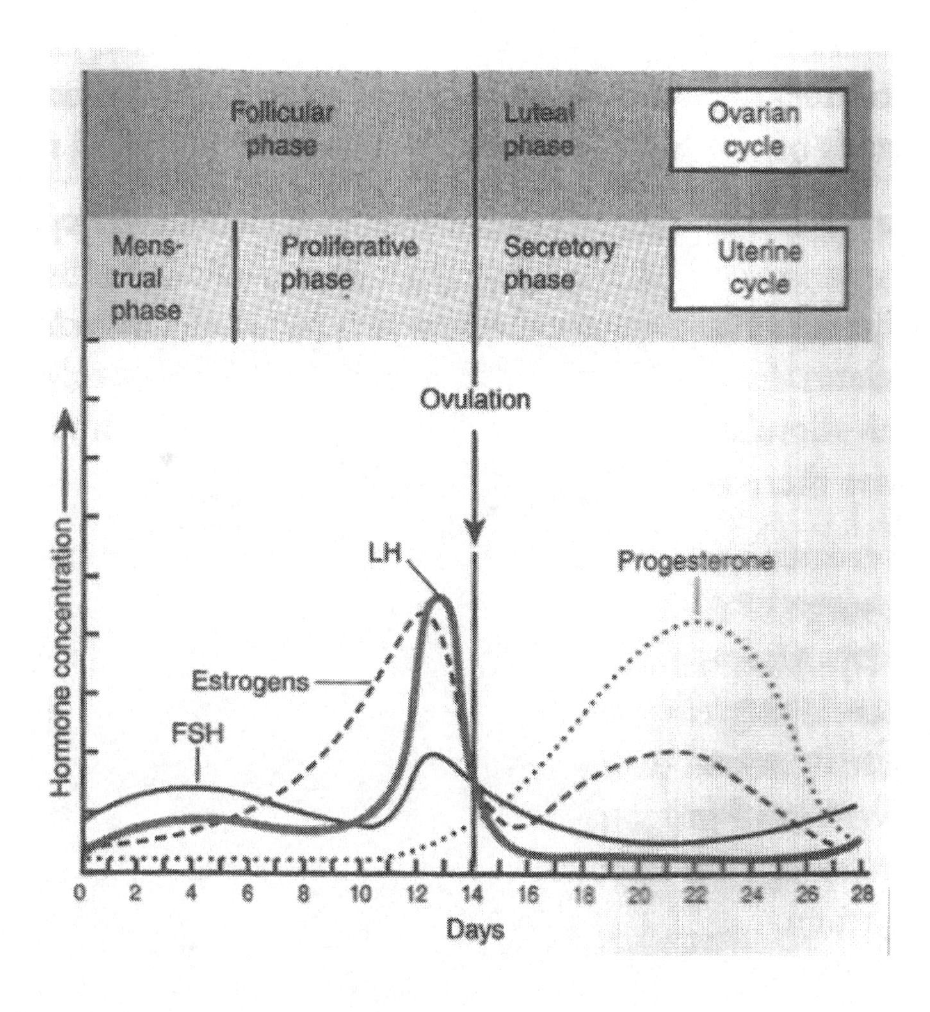

For accurate determination of the ovulation date, you need two methods. An ovulation kit, and body temperature monitoring. My recommendation is to use both. The reason is that temperature monitoring, unless it is done more than once a day, can be missed.

And ovulation kits detect a hormone that peaks first just before ovulation. The time gap in between the LH peak and ovulation may differ based on individual cases.

The LH approach.

This is my first recommendation. Ovulation detection kit.

There are a few good test kits on the market that would detect ovulation date based on the fact that Luteinizing Hormone "LH" peak is what triggers ovulation, in most women within that day.

This chart is based on the fact that, as with all our physiological functions, the brain glands control this operation via different hormones.

In this case, the hypothalamus and pituitary glands control the ovulation hormones. Luteinizing hormone LH stimulates the release of the egg from the follicle into the fallopian tube where pregnancy would take place.

I recommend you would, simultaneously, use at least two determination methods. The results should compliment or confirm one another.

Using this approach, test three times daily. You should practice testing monthly for a few months prior to the "real thing."

Now, in addition to LH, as a supportive measure, use the body temperature record approach.

Here is how:

Temperature record.

This is the recommended supportive test.

Wait until bleeding days are finished, start with the 6th or 7th day of the cycle (1st day is when bleeding first appeared.)
For consistency, temperature should always be taken first thing in the morning as soon as you wake up, before getting out of bed. It could also be taken at night in bed after relaxing in bed for at least an hour.

The normal oral temperature is around 37C (36.1 to 37.8).
Use a chart paper, available at stationary stores and assign a vertical increase of 0.1 C for each graduation. While the horizontal indication would be the dates. Assign 15 days, starting from the 7th day.

Here is an example:

Actual record submitted by a friend I assisted, and it is utilized here as an illustrative sample. Individual cases may vary slightly.

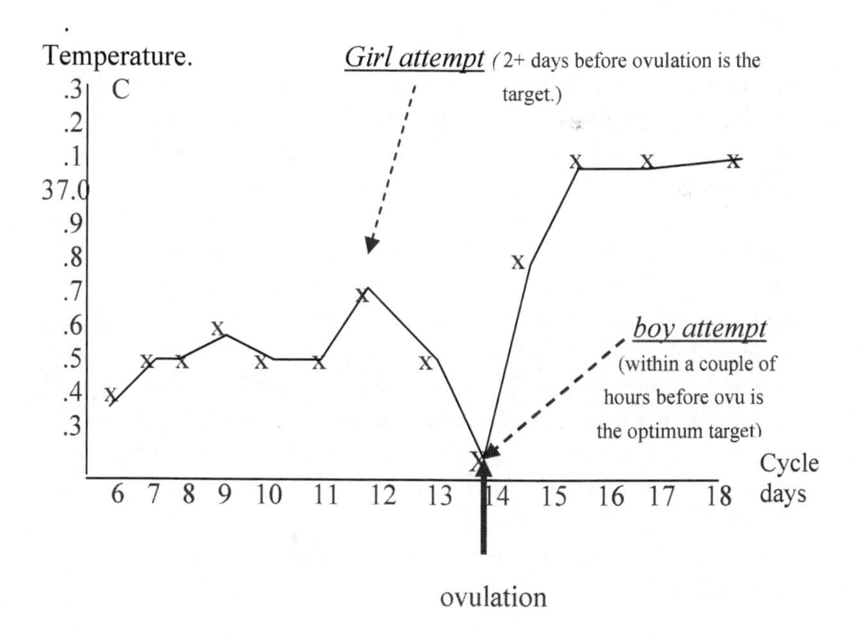

The noticeable drop in temperature due to the drop in estrogen, and the rise of the temperature with the start of a progesterone peak, is the ovulation time.

Here is a typical textbook basal temperature graph you would find in many magazines and books.

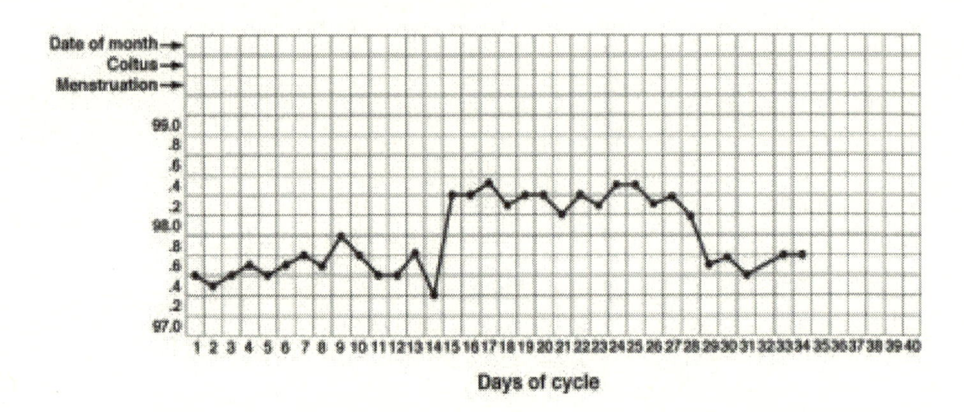

Using the temperature approach can be beneficial when using the LH kit. The reason is that LH peak, which is what the test detects, happens first, then ovulation. Depending on the individual case, one could not be certain as to how short (an hour?) or how long (over a day?) the gap is between the LH peak and ovulation. This is significant because for maximizing the chance for a boy, the egg should have just been released, or is being released. This means that the Y-sperm will not have to wait for ovulation. The waiting period will present double jeopardy: Y-sperm don't live as long as the girl-producing X-sperm and are not as hardy. And, a waiting period will give the slower X-sperm the chance to catch up and to compete for the egg.

Having said that, after ovulation the vaginal secretions become acidic again.

The egg waits for only about a day, feels stood-up, and leaves with her servants of cells "the period discharge." Once that happens, it's too late. Try again the following month.

A health magazine that was brought to my attention addressed this issue from a different angle. A sex-selection Dutch scientist wrote about the ovulation symptom known as "mittelschmerz," which is discussed by several European papers. Simply, a lower abdominal discomfort often felt, but dismissed by many women.

Some may feel it as a minor aching, others as a sharp painful nudge in the right side of the lower abdomen at the time of ovulation.

If you listen to your body carefully, and you practice for a few months – along with using the supportive LH kit and temperature record – you're likely to feel that exact moment of ovulation.

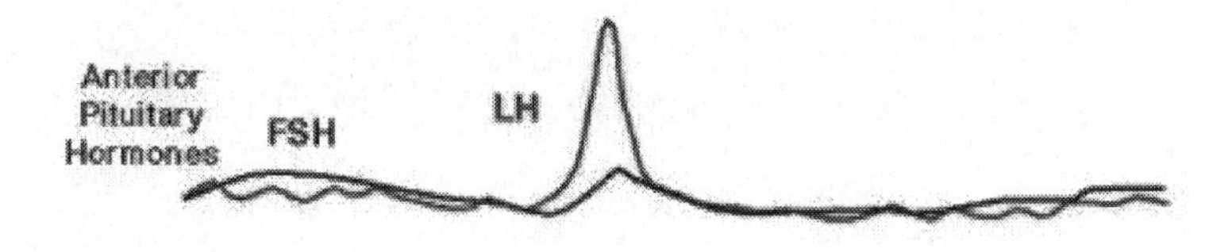

As shown above, kits make use of the LH peak to detect ovulation.

I do not recommend depending on methods such as monitoring the consistency of the cervical mucus. Yes, it would give the woman a good indication as to when she ovulates. She would need to probe, using the middle finger, for a sample, inserting it deep until she could feel the cervix, and obtain a sample. The mucus increases as the woman gets closer to ovulation, and most importantly, it gets thinner to create an easier travel environment for the sperm. It also becomes clear and slippery.

This is, however, more of a guideline approach. Not good enough on its own. Stick with the temperature approach as a supportive measure to the ovulation kit test.

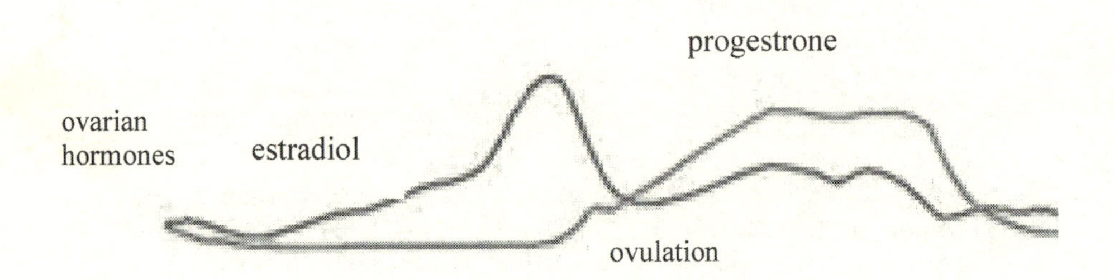

As indicated on my chart a couple of pages back, if aiming for a boy, plan intercourse between the drop in temperature and the peak high, within +/- a couple of hours margin.

If the plan is a girl, 2 or 3 days before ovulation is the optimum target.

In either case, it is obvious that accurate determination of ovulation date is essential.

Synopsis:

- Use an ovulation detection kit to determine the time ovulation is about to take place. It detects the LH peak that triggers ovulation, not the exact moment of ovulation itself.

- Use also the body temperature record approach. Ovulation is the gap between the drop and the peak of the graph.

- Use both methods simultaneously. In addition to the reason explained by the preceding page, some women are best suited for one test than the other. e.g. those with an unusually short cycle would be better suited for the use of ovulation kits.

- For a boy, almost the exact ovulation time is the target.

- For a girl, two days prior to ovulation would be the target for intercourse.

- Ensure that you are healthy. Three scientific reports suggested that an infection could influence the vaginal milieu.

Notes:

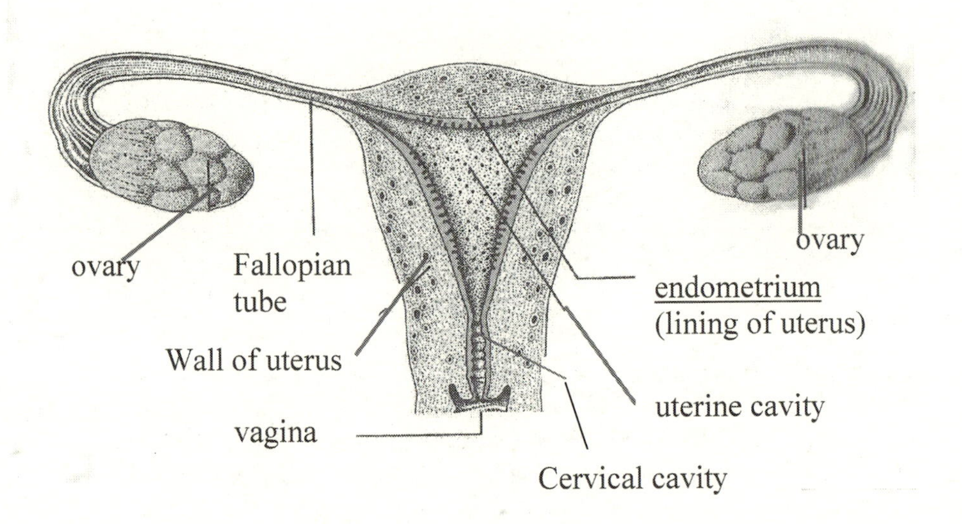
ovary
Fallopian tube
ovary
endometrium
(lining of uterus)
Wall of uterus
uterine cavity
vagina
Cervical cavity

A mammal is a mammal.
Well, almost.

This is a chapter the typical couple could skip.

As I said, this applies in general, with minor alterations, to all mammals. When I bred my Great Danes, I did the AI "artificial insemination" myself. The sire was allowed to mount the dam after placing a collection bag between them, and the AI procedure would allow for a deep deposit, as deep 11-12 inches insertion, or as shallow as 4-5 inches insertion - depending on whether I was aiming for males or females. But, the second part of the procedure was the most important. Separating all (likely most) of the Y sperm from the X sperm before the AI insertion. We'll talk about that later.

The reproductive system of dogs and humans are very similar. In the female, the reproductive system is composed of the ovaries, oviducts, uterus, cervix, and vagina. The ovaries are the site of production of the unfertilized eggs, and many of the hormones responsible for heat cycles and the maintenance of pregnancy.

The eggs pass from the ovaries into the oviducts. These small finger-like tubes are the site of fertilization by the sperm. From there the eggs pass into the uterus, which is composed of the left and right horn and uterine body. The developing embryos mature within the uterus, attached to its walls by the placenta which also surrounds them.

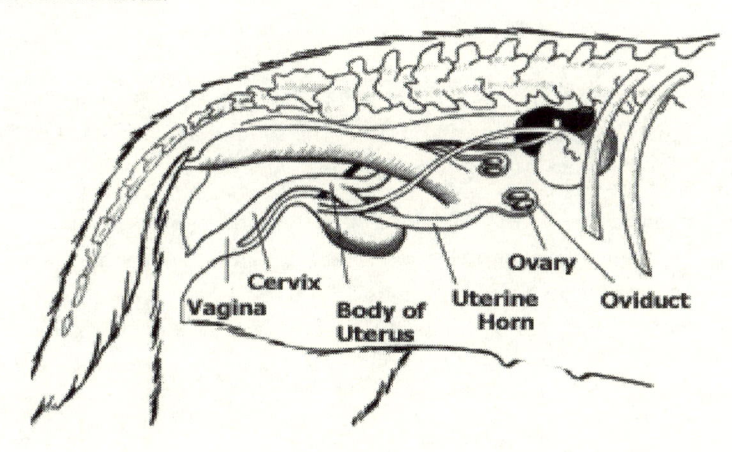

Note here that one would have to guide the insertion tube with the longer middle finger, up into the vagina, to the cervix, (since the same opening leads downwards into the urinary system.)

A sterilized syringe, a collection bag, and a flexible insertion (insemination) tube are used, and the sperm is inserted right after collection (and conditioning as I'll explain) - into the female.

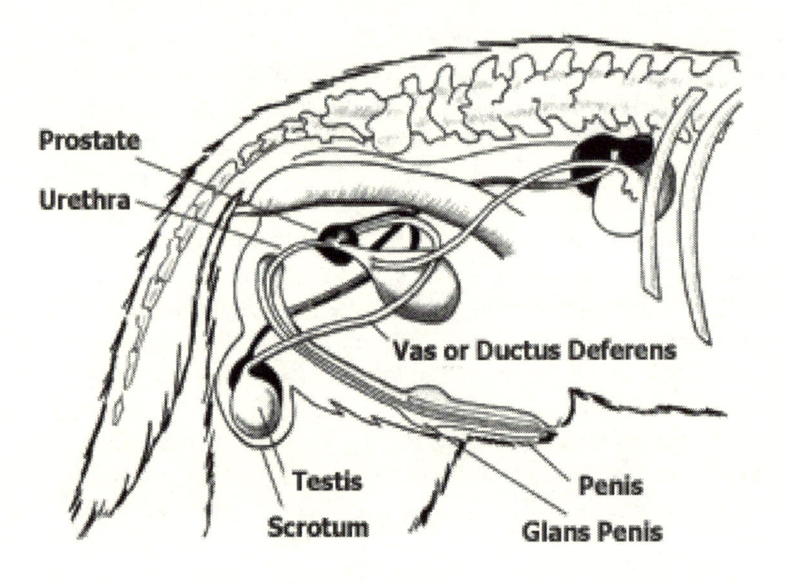

We'll talk later about A.I. for human application. (Needless to say, do not use a condom to collect the semen, as they are coated with spermicides to kill the sperm.)

We'll also address the enrichment of ovulation and how in animal breeding, it can be used to induce the female's estrus (bring her into heat on command.)

There are multiple hormones that help regulate the estrus (heat cycle) and pregnancy in mammals, including dogs.

Estrogen: It stimulates the ovaries to produce eggs.
Luteinizing Hormone (LH): It stimulates the ovaries to release the eggs.
Progesterone: It maintains the pregnancy.

Understanding this helps in determining the best time to breed.

Most mammals ovulate when the estrogen level in the blood is increasing. Dogs, however, ovulate when the estrogen level is declining and the progesterone level is increasing. Estrogen levels can give us a general idea of when a dog comes into heat, but are not sufficient to determine when breeding should actually take place.

Vaginal cytology can also provide some general information. Progesterone levels, and luteinizing hormone (LH) levels are the best indicators of when ovulation takes place and when is the best time to breed. They are also useful in determining whelping dates, allowing an owner to reserve the appropriate days on the calendar and even to schedule a c-section (caesarean) weeks in advance.

Luteinizing hormone level

LH is species-specific, meaning it is chemically different in different species. Blood testing for LH, therefore, needs to be done at a vet lab.

The LH test needs to be done daily starting toward the end of proestrus. The LH spike typically lasts only 20-26 hours. Hence the test must be done daily to spot it. The peak usually matures 48 hours prior to ovulation.

The following chart illustrate comparative hormonal levels.

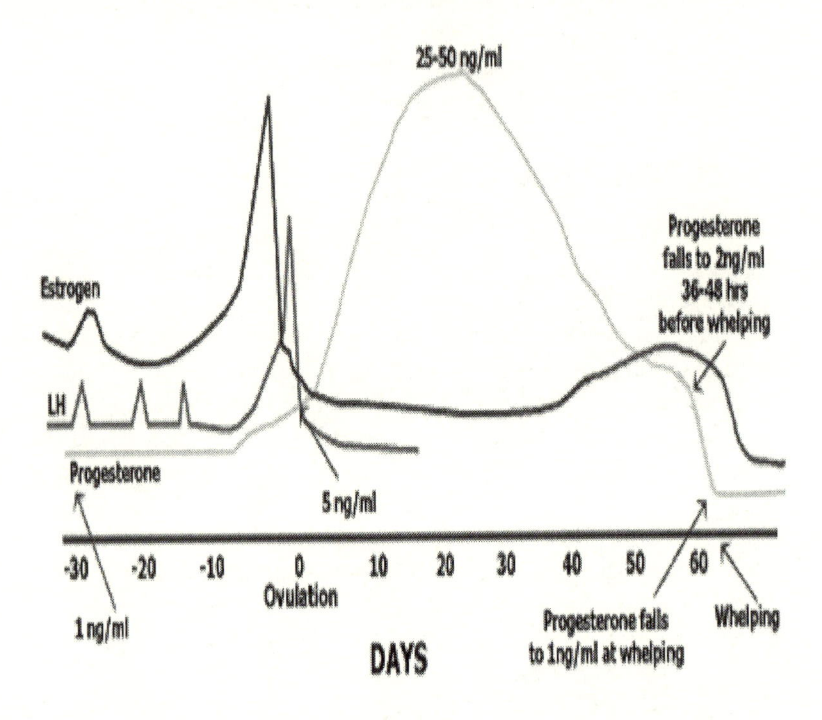

Progesterone levels and breeding

The progesterone test is not species specific, i.e. the test can be performed in either a human lab or a veterinary lab.
A progesterone test is needed every 2-3 days starting about 4 days into the heat. Timing of the test can be more certain if the lengths of the dog's previous heat cycles were recorded.
The beginning progesterone levels are typically less than 1.0 ng/ml until the day before the LH surge. The day of the LH spike, serum progesterone concentrations are 2-3 ng/ml.

The day following the LH surge, the serum progesterone concentration is 3-4 ng/ml. Ovulation occurs on a progesterone level of 5 ng/ml.

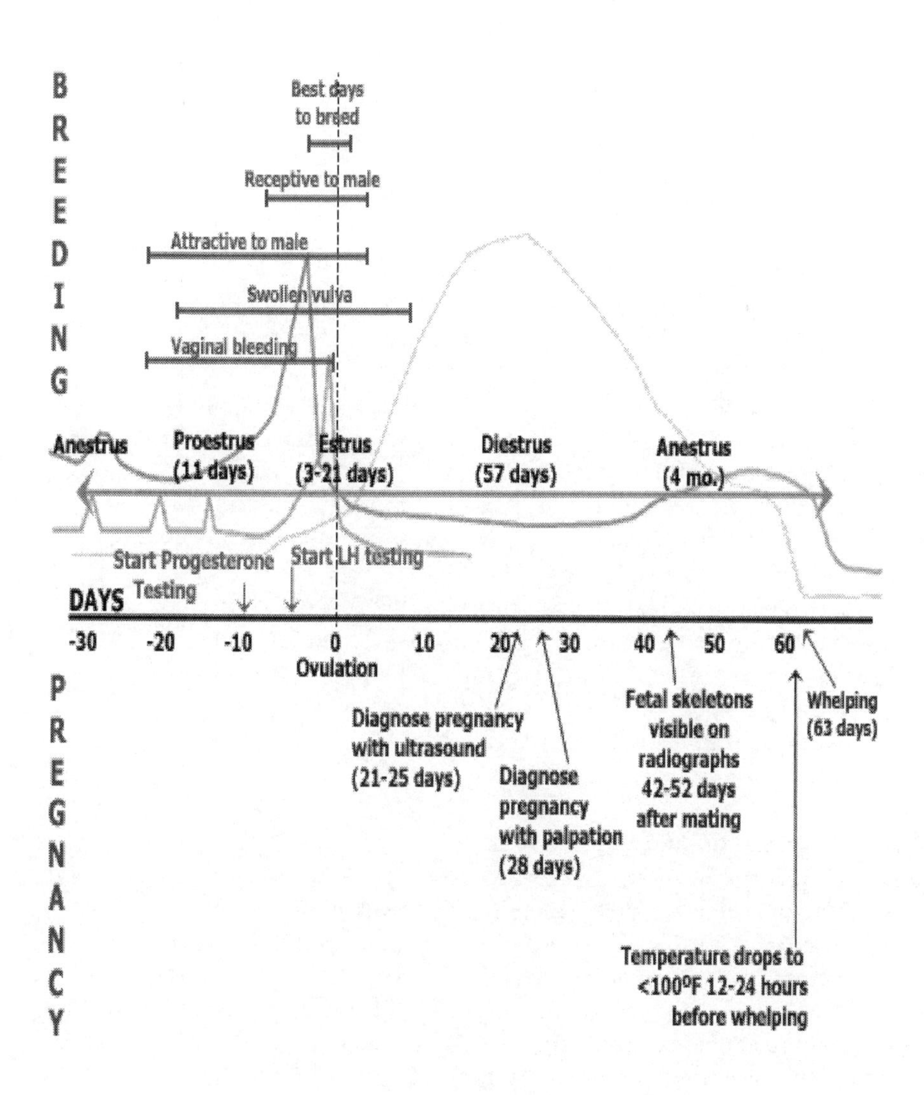

Timing of breeding

The aim is to identify when the progesterone level reaches 2.5 ng/ml so the mating schedule can be set up, or the veterinarian and owner of the male dog can be notified for the stud's semen collection. Depending upon the type of semen used, optimal times for natural or artificial insemination are:

Natural breeding is not the goal here, since we are targeting sex selection. Otherwise, it should occur 3 days after the 2.5 ng/ml mark. Sperm in fresh semen survives about 5 days after insemination.

Artificial insemination using fresh chilled semen should be used for a 1-time breeding. Insemination should take place 4 days after the progesterone reaches the 2.5 ng/ml mark or 48 hours after the 5 ng/ml mark. Sperm in chilled semen survives about 48 hours after insemination. With artificial insemination, the semen should be deposited into the cervix to increase the chance of it being drawn into the uterus. Pressing against the inside of the vaginal opening, an inch deep, will simulate the natural "tie" and will trigger her muscular contraction pushing the sperm forward.

Artificial insemination using frozen semen should be performed 5 days following the 2.5 ng/ml mark or 72 hours after the 5 ng/ml mark. Sperm in frozen semen survives less than 24 hours after insemination. Frozen semen is ideally deposited directly into the uterus through surgery.

Fertilization and implantation

The sex selection rules are a little different here.
The sperm requires a period of a few hours after ejaculation before they are capable of fertilizing an egg. This period is referred to as the "capacitation time." The egg also needs time to mature after it is ovulated, generally 48 hours from ovulation until it can be fertilized. Fertilization occurs in the oviduct (Fallopian tubes) regardless of the method of insemination. The

fertilized egg then travels into the uterus but does not implant until 17-18 days after ovulation. If there are problems with the lining of the uterus, the egg may not implant or the placenta may not develop further. A normal placenta grows into the lining of the uterus.

Progesterone levels during pregnancy and whelping.

After ovulation, progesterone concentrations continue to increase for 2-3 weeks, finally reaching 10-80 ng/ml. This level is necessary to maintain pregnancy. In the dog, the progesterone level will remain at this level for about 60 days whether or not the dog is bred, and whether or not she is pregnant.

About 48 hours before whelping, the progesterone level drops to the 2 ng/ml range and within about 24 hours of whelping, the level drops to the 1 ng/ml range.

This can help determine the time of whelping, especially if the progesterone level or LH level were not used to determine the ovulation date.

Heat cycles

The heat cycles of the female (dam) are the result of, and are controlled by, hormones that are produced and released by the ovaries and other glandular structures within the body.

The ovaries are paired structures that become increasingly active when the mammal passes through puberty.

This ranges from between five and twelve months of age depending on the "line" and the size of the animal. In the small breeds, heat cycles occur as early as five months of age, while in the giant breeds, this may not occur until the animal is about a year-old. Typically, these cycles will occur every six to nine months throughout the life of the animal. Dogs do not undergo any form of menopause. There have been rare cases of heat cycles resulting in pregnancies at fifteen years of age.

The estrus cycle of the female is divided into four different stages. There is great variation in the length of these cycles among individuals of the same breeds and among various different breeds. Additionally, the same animal may have significant variations over the course of a lifetime. It is therefore, impossible to talk about the cycling of bitches using exact dates or time periods.

Proestrus: The first stage of a heat cycle is a preparatory period referred to as proestrus. This follows a period in which the reproductive system was from all outward appearances, inactive. Proestrus typically lasts five to nine days. On the first day of proestrus, the vagina becomes swollen and a bloody discharge is soon observed. During this stage, males show an interest in the female, but she will be unreceptive to them.

Estrus: The next stage is referred to as estrus. This is the active breeding phase, and will usually last from five to nine days. Bleeding from the vagina is very slight or completely absent at this point. Eggs are released from the ovary and travel down the oviduct. During estrus males will attempt to mate with the female. The female will allow them to mount her, resulting in intercourse. In the dog, a 'tie' usually occurs in which the male and female are held together physically, with the vagina tightly enclosed around the glans penis. Ejaculation will occur, and sperm cells will enter the uterus and make their way to the oviduct where their union with the egg will result in the formation of a fertilized egg which is referred to as a zygote. This matures further, developing into an embryo and then a fetus.

Diestrus: Following estrus is the diestrus period. This extends from the time when the female dog is no longer receptive to the male to the end of pregnancy. In cycles in which a pregnancy did not occur, diestrus will last for a period of up to 80 days.

In early diestrus, the embryos and their placentas attach to the wall of the uterus, from which they will derive their oxygen and nutrients.

Anestrus: Following diestrus is anestrus. This is the quiescent period between heat cycles characterized by no outward physical or behavioral signs of sexuality.

In the male, the important structures of the reproductive system are the testes, ductus or vas deferens, prostate gland, and penis. Sperm production and storage occur within the testicles. Upon ejaculation, the sperm is transported to the prostate gland by the vas deferens. Within the prostate, additional fluids are added to the sperm to nourish it and to transport it through the uterus.

The sperm and prostatic fluids, at the level of the prostate gland, enter the common urethra and are carried from the body through the penis. The penis of the dog has two specialized structures.

The glans penis is a bulb-like dilation at the base of the penis, which fills with blood and holds the penis within the vagina during intercourse. Within the penis is a bone that maintains the shape and direction of this organ during mating. The penis is protected from the environment, as it is enclosed within the sheath or prepuce.

Understanding the preceding basics is a must before one should attempt to breed via CAI in order to choose the sex of the puppies.

CAI, conditioning or treatment of the sperm, including simple sperm separation techniques that'll be explained in the next section, are a must when breeding for a specific gender.

Part – II

Diet.

Artificial Insemination "A.I."

Conditioned A. I.

Sperm separation.

Diet -
The underestimated element.

The first section of this book ended with the coverage of the basics, the facts about the sperm characteristic and elements related to a woman's body.

Diet's role in sex determination has been around for centuries. The 1898 Schenk's book wasn't the first material or study on the subject. Even the famous 1930 Herbst book wasn't the first to announce the specific effect of K, Ca, Mg, Na ratio on sex determination. Though I haven't read either book (only read about them) – I have reviewed other papers and articles that were written before these books were published.

The following section of this book elevates the sex selection to a much higher peak that provides more than a very good chance of success.
It very well guarantees it.
It's the use of:

- **Diet or physiological or specific chemical conditioning.**
- **Conditioned artificial insemination.**
- **Sperm sorting.**

We'll start with diet.

In breeding other mammals: cattle, horses and Great Danes experience, I developed some techniques and formulas, and I was able later, with the assistance of selected associates of biochemists and a gynecologist working for O.G.L.C. under my guidance, to perfect them for use with a cousin mammal, humans.

Let me start with mammals that are not human:

- A diet for the male to promote a higher sperm count.

- A diet for the male to promote Y-sperm success. It should be noted that Y & X spermatozoa are not generally of equal numbers. Sperm morphology is consistent with several dietary factors influencing sex-determination genotypes.

- A diet for the female to promote healthy ovulation. A twist to this ovulation enrichment can trigger ovulation sooner, or could, in human applications, produce two eggs – depending on the nature and objective of the ovulation enrichment programme (this will also be addressed in the technical paper.)

In other mammals – a recent experiment "on Great Danes in particular" was successful in inducing her estrus "bring her into heat" within 3 weeks.

> A.I. is a must in sex selection for mammals other than humans. Manipulating the female's cycle might be a preferred option. At times, I preferred to do the artificial insemination in late winter or early spring, avoiding whelping of the puppies in the harsh winter or the hot summer months, allowing the pups to move to the outside kennel in the late spring or early summer.
>
> Now, all mammals share these facts and factors:
>
> - Diet influences the zona pellucida, a translucent protein coating the ovum. This is a major element in the fertilization process that can, with diet (the body's chemical manipulation) be used in favour of a specific spermatozoa.

- Diet influences the composition of the cervical mucus, which is a colloid, a hydrophilic gel with several electrolytes and organic compounds including amino acids, soluble proteins and glucose. Cervical mucous composition can favour one sperm over another. It can also provide the fast but weaker sperm Y with a sprinter-like energy to make it through the ciliated cells lining the fallopian tube. Diet will also influence the performance of anti-peristaltic muscular contractions in assisting sperm in general towards the egg. With the intelligent manipulation of sperm characteristics, sex selection is made much more achievable.

- Diet influences the sperm capacitation, the sperm's removal of an enzyme coating, a necessary act for penetration of the ovum.

The diets designed to improve the overall sperm health and count, are designed for all mammals, men included. The same goes for the diet structured to give the Y-sperm a helping hand. I'll explain later.

The critical issue here is the need to test the individual candidate's level of several elements such as glucose, blood pressure and even K, Ca, Mg and Na before an individualized, customized diet can be optimized for effectiveness. Pushing the limit on Ca-Mg-K-Na ratio will be a crucial factor. For this reason, I did not provide a specific "menu" to follow. If your K level is too high or too low, the ratio we mentioned will be off. Hence, you'd need to work closely with your health consultant on a personal level.

I'll present the information in a point format to minimize the need for connective sentences, and to make for an easy reference. Here are selected tips of DOs & DON'Ts presented along with necessary selected health factors. Though some medical clarifications will be necessary to understand the whole picture, the technical part is presented in a reasonably simple language.

- The male's diet for at least a month prior to intercourse must enrich the levels of: zinc, vitamin C and E.

- A mixture of: a cup of strong regular tea, a cup of strong regular coffee (not made with dark beans. Contrary to the common belief, the darker the coffee bean, the less caffeine) - mixed with a generous amount of "pure" chocolate and some honey (chocolate is limited to human consumption only) – is a high caffeine dose that will stimulate the sperm in general. However, since the Y-sperm is faster, it will have a better advantage in the early stage of the race, and, with a deeper deposit shortening its distance from the egg, its advantage is even maximized. This dose is administered twice, first 2 hours, then half an hour prior to intercourse when trying for a boy.

- Organic dairy is the best choice of dairy sources in general. However, dairy foods such as milk and cheese may be congesting to the body. In cases of congesting fertility issues such as PCOS and Endometriosis, they may aggravate the imbalance. Observe how your body reacts.

Dairy that is not organic should be avoided as it contains added hormones and antibiotics which can contribute to increased estrogen levels within the body. There are many healthy alternatives to dairy such as fresh almond or hemp milks.

- Omega 3 in fish is an essential fatty acid. These fatty acids aid in the production of hormones, reduce inflammation, and help regulate the menstrual cycle. A diet very rich with fatty acids or fish-based for two months prior to the intended pregnancy attempt is advisable. Fish is also a great source of protein and vitamin A. Avoid large deep water fish such as Chilean sea bass, tuna and swordfish as they often have high level of mercury. A safer choice would be cold-water fish such as cod, wild Alaskan salmon, and halibut. Also when choosing salmon, avoid north Atlantic farmed salmon and choose wild salmon instead. Farmed salmon contains antibiotics and toxic food dyes.

- Choose meat that is grass-fed and organic. Conventionally raised cattle contain high levels of added hormones and antibiotics which also can contribute to estrogen dominate conditions. If you are experiencing endometriosis, reduce the amount of red meat consumption..

- Eat only grains in their whole, natural form. They are filled with fiber, important vitamins, and immunity supporting properties. Fiber is important for helping the body to get rid of excess hormones and it helps to keep the blood sugar balanced. Avoid processed and refined white foods and grains such as white bread, semolina pastas, and white rice. Instead choose whole-wheat or sprouted bread, rice or whole-wheat pasta, quinoa, and brown rice.

- Fiber helps to regulate the blood sugar levels which in turn helps to reduce fertility issues such as PCOS and immunological issues. It also promotes healthy hormonal balance. Some examples of high-fiber foods are fruits, vegetables, dark leafy greens, and beans.

- To conceive a boy, both parents' diet – but by far, more importantly the mother, must increase potassium and salt (sodium chloride) and lower calcium and magnesium.

- To conceive a girl, both parents' diet – but more importantly the mother, must increase calcium and magnesium and lower potassium and almost eliminate table salt.

- Sodium in high salt diets has been associated with the conception of boys. A century ago, a chemist who retired to run a farm gave his cows sodium chloride lumps to lick and succeeded in producing 88% males. A Canadian doctor in the 1960s experimented with near 300 women. Those who wanted a boy were placed on a high potassium, high salt diet. Those who wanted a girl were placed on low salt, low potassium, high-calcium diet. The results were published in the international journal of gynecology and obstetrics, announcing an astonishing 84% success.

- For a boy, the daily diet should provide a guideline of:

250 mg Ca 100 mg of Mg 4500 mg K 5000 mg Na

Potassium chloride tablets and sodium chloride (table salt) are easy to absorb and are a good source of K & Na.

- For a girl, the daily diet should provide:

2000 mg Ca 400 mg Mg 600 mg Na 1500 mg K

A similar published study in both Germany and France, using this guideline diet, reported an average success rate of 86%. This is incredible indeed, and it's higher than any success rate reported by merely depending "solely" on sperm characteristics techniques.

- Soy foods, unless fermented, have been shown to contain estrogen mimicking properties, causing a negative impact on your hormonal balance.

- Endocrine, toxins in our environment, known as xenoestrogens, bind to our estrogen receptor sites, producing an estrogenic effect. As well, phytoestrogens, compounds found in plants, also have the ability to bind to estrogen receptor sites and mimic our natural endogenous estrogens. Phytoestrogens have been shown to have a weaker estrogenic effect than our own endogenous estrogens.

Xenoestrogens's ability to bind to our estrogen receptor sites disrupts the function of the endocrine system. Not only can they mimic our natural hormones, but they can also block other hormones from binding to receptor sites. They endocrine disruptors can alter how natural hormones are produced, metabolized and eliminated.

Common known substances that have demonstrated estrogen-mimicking effects on animals (including humans):

* Atrazine (weed killer)

* Ethinylestradiol (a comibined oral contraceptive pill, released into the environment as a xenoestrogen via the urine and feces of women who use this contraception.)

* Butylated hydroxyanisole known as BHA (food preservative)

* Methylbenzylidene camphor known as 4-MBC (suncreen lotions)

* Bisphenol A (found in polycarbonate plastic and epoxy resin)

* Heptachlor and dieldrin, DDT (insecticides)

* Erythrosine, FD&C Red No. 3 (food dye)

- Endometriosis is generally caused by:

* Estrogen is too high, or progesterone is too low.
* An abnormality in the immune system.
* Consumption of excess meat.
* Radiation and EMFs.

Endometriosis results in excess tissues growing in the uterus (as well as other areas of the body) which makes it harder for an embryo to attach and grow healthy. As these excess tissues bleed, inflammation develops scar tissue, and this process results in attachment to the uterus, fallopian tubes, the ovaries, and other organs, causing “congestion." In addition, the woman’s body may form antibodies against the misplaced endometrial tissue. These antibodies may even attack the uterine lining and cause miscarriages – however, this is an issue beyond the elements of the impregnation control subject of this book.

The right diet is very important for healthy ovulation (and pregnancy). Fiber helps the body to get rid of excess estrogens.

Some good sources of fiber are:

- Dark leafy greens
- Quinoa
- Chia seeds
- Ezekial bread products
- Broccoli
- Swiss chard

- Phytoestrogens are not true estrogen, but they may have a similar action to our own endogenous estrogens.
The balance between their necessity and their danger is often a thin line.
Phytoestrogens exert weaker estrogenic effects on cells, compared to endogenous estrogens, or xenoestrogens.

As indicated, phytoestrogens have been shown to have an anti-estrogenic effect premenopausally. They do this by competing for hormone receptor sites. If they bind to the estrogen receptor sites first, they actively block xenohormones from binding. This occupies the receptor sites with less estrogenic phytoestrogens, resulting in the protection of the receptor sites from considerably stronger xenohormones.

Phytoestrogens types, in order of strength, are:

1. Isoflavones are most commonly found in soy, isoflavones, red clover, alfalfa and the legume family.

Note that soy products in the diet contain concentrated forms of phytoestrogens, which may lead to estrogen dominance.

2. Lignans- commonly found in flax and sesame seeds.

3. Flavanoids- powerful antioxidants and anti-inflammatory found in most yellow, purple, red and black fruits, and in Royal Jelly. They weaken estrogenic effects.

4. Coumestrol- found in high concentrations of sprouted soy and red clover.

- It is important to know that in order for phytoestrogens to be metabolized by the body, healthy intestinal flora is essential. You need to also consume healthy pre-biotics, and probiotics such as fermented foods (e.g. yogurt, kefir, kimchi, saurkraut, kombucha.)

Prebiotics help to establish and promote the growth of good bacteria (probiotics.)
Prebiotics act as food for the probiotics in the digestive system. Prebiotics: soluble fibre is an important source of prebiotics, so aiming for the minimum 25g of fibre per day for adults will help you to get more prebiotics in your diet. Inulin is a naturally occurring prebiotic that is found in vegetables, grains, and roots, such as chicory root.

Probiotics: 1 billion live bacteria per serving is ideal and can easily be included in your daily intake. Other examples of probiotic food are sauerkraut (fermented cabbage), kimchi (Korean spicy cabbage), tempeh (a fermented soybean product), and soy sauce.

- Several pharmaceutical compounds have utilized intelligent molecular design. Also pharmaceutical extractions from herbs and other valuable plants are a source we must not ignore.

Some elements our body needs beyond basic food, are necessary for optimum health. In fact (and I know this is another subject) – all the vegetables and fruits one could eat on a day are not sufficient when it comes a few specific vitamins and minerals our body needs. Fruits and veggies, in some cases, will only provide you with the equivalence of 10 to 50 units - depending on which vitamin or mineral. Yet, depending on your condition and lifestyle, an intelligent supplementary diet requires 700 to 1200 units of such vitamin or mineral daily.

- Diindolymethane, known as DIM is an example I examined along with two fellow chemists. It is shown to promote a healthy estrogen balance. DIM comes from the plant chemical I3C, short for indole-3-carbinol. DIM stimulates the body to metabolize "bad" estrogens (not your good ones.) It is not, however, anywhere near being sufficient alone, but it does address one part of the complete picture.

Note that in order for estrogen to be modified into its final form before "body ejection," it has to be combined with oxygen for aerobic metabolism. DIM increases specific aerobic metabolism for estrogen, which in turn, multiplies the chance for estrogen to be broken down into good beneficial estrogen metabolites (2-hydroxy estrogen.)

When the good estrogen metabolites are increased by the DIM, a reduction in bad estrogen (16-hydroxy estrogen) is produced, which presents the threat of endometriosis, uterine fibroids and even cancer. While "good" estrogen metabolites provide us with the protection of antioxidant activity, elevated "bad" estrogen may be promoted by obesity and exposure to man-made environmental toxins.

- Maca is another example. It is an adaptogen herb that supports the pituitary, adrenal, and thyroid glands, promoting healthy endocrine system function. It helps regulate estrogen level in women. As indicated, excess estrogen levels also cause progesterone levels to become too low. It is shown to help increasing the progesterone levels which are essential to carrying a healthy pregnancy.

Maca is also part of the total supplementary formula for men. It has been shown to increase libido and healthy sperm.

- Pharmaceutical analytical chemistry utilized herbs to extract and formulate women supplements helpful for such issues as endometriosis – and to support the body in eliminating excess hormones, endometrial tissues, and inflammation, while promoting the reduction in endometrial growth. Of course, we do not support the blind, uneducated use of herbs as found in the wild. Though they have helpful substances, they also have harmful substances ignored by those who promote, for one reason or another, raw herbal mixtures.

- A major part of increasing egg health and preparing the uterine lining is to take a prenatal multivitamin. Making sure your body has all of the nutrients necessary is a lot easier when you are taking a multivitamin. However, it is imperative to understand that multi vitamins and minerals for women, the kind of one size fits all, is a dangerous game to play. You must obtain a personal profile via your doctor, with a complete analysis showing what is low, what's high, what will affect what, etc.

A customized personalized profile is important to both men and women in all age categories.

Foods for egg health:

-Berries
-Dark Leafy veggies
-Halibut
-Salmon
-Maca
-Broccoli
-UltraGreens

-Pumpkin seeds
-Ginger
-Sesame seeds
-Royal Jelly
-Turmeric

Foods that damage egg health:

-Sugar
-Cigarettes
-Caffeine
-Alcohol
-Low fat diet
-Processed foods
-Trans Fats
-Non-organic meats and dairy
-Soda

- Antioxidants are one of the most important components to having healthy fertility for both men and women. Antioxidants help to protect the egg and sperm from free-radical damage. Free radicals are able to damage both cell health and the cell's DNA. Women were born with the initial form of the ova (eggs). And being old cells, they are vulnerable to free radicals.

- L-arginine, an amino acid, has been shown to increase ovarian response and endometrial receptivity.

Fertility superfoods are extremely nutrient dense and deliver therapeutic benefits every time you take them.

I mentioned Maca as an incredible fertility super food. It helps to balance hormones, though it does not contain any hormones itself.

Royal Jelly is a fertility superfood. It helps to increase the egg quality and quantities. Royal Jelly is the food that only the queen bee eats.

In experimenting on mammals other than humans, I adopted a full programme combining supplements and physiological conditioning. I have never failed in bringing my Great Dane dams into estrus whenever I wanted, within 3-5 weeks. Of course general health prior to that specific programme is an element.

The physiological aspect of the programme is similar to and is supported by another experiment conducted on humans in the Netherlands. In a study of women who were not ovulating, one group received cognitive behavior therapy, and the other group was just observed. 84% of the women who received cognitive behavior therapy resumed ovulation again, as apposed to only 19% from the randomized observation group.

- Several oils are part of the pre-ovulation natural treatment.

Geranium (Pelargonium gravolens)
An adrenal cortex stimulant; helping to regulate and balance hormones. It also helps to detoxify the lymphatic system and helps to alleviate anxiety.

German Chamomile (Matricaria chamomilla, Matricaria recutita) promotes calming of the nerves.

Rose Otto (Rosa centifolia, Rosa damascena, Rosa gallica): helpful in treating PMS, regulation of the menstrual cycle and has been shown helpful for women who have trouble conceiving. It relaxes the uterus. It is an aphrodisiac, helping to inspire confidence in expressing one's elf sexually.

Sweet Marjoram (Origanum marjorana, Marjorana hortensis) is helpful with menstrual cramps. However, it has been shown to reduce libido, so it may be best used only during menstruation.

Jasmine
(Jasminum grandiflorum, Jasmine officianale)
Jasmine is a wonderful oil as it aids in overall female reproductive health, but it also strengthens the male sex organs.

Sweet Fennel (Foeniculum vulgare, Foeniculum officinale, Anthum foeniculum):
Helps to regulate the menstrual cycle, may help reduce hormone fluctuation.

Cypress (Cupressus simpervirens):
This oil is great for women with over-heavy menstrual bleeding and painful menstruation. It is reported to be very astringent.

This blend of oils will increase women libido: Ylang, Sandalwood, Rose Otto, Jasmine, Patchouli and Clary Sage.

This blend has strong aphrodisiac qualities, and is suitable to massage on women when trying for a boy, to assist your woman to achieve orgasm for the reason explained in a previous chapter.

- Protodioscin, a steroidal saponin compound in the Tribulus, Trigonella and Dioscorea families, aids in fertility for men. It improves DHEA levels in the male body where ED has been found to be associated with low levels of DHEA.

- Antisperm Antibodies are triggered via the immune response that works to kill off sperm. High numbers of sperm antibodies can make it difficult for the sperm to reach the egg, and can make it hard to achieve a boy.

Antisperm antibodies also may damage sperm that survives, which increases the chance of a miscarriage, and, it has been associated, for reasons explained, with carrying a girl.

- Tribulus Terrestris: Fertility Herb for Both Men and Women.

Tribulus has been successfully used for Female Fertility treatment and "Ovulation Stimulation."

A study performed on 57 women who were not ovulating showed that 77% realized normal ovulation after only 2 months of treatment with a 500 mg daily dose from day 5-14 of their menstrual cycle. 11% became pregnant right away.

A rat study using Tribulus, published in Aug. 2011, showed that Tribulus reduced the number of cysts in the ovaries in female rats with PCOS. High doses of the extract were administered orally. The treatment showed the ovarian cysts to have significantly decreased, and normal ovarian function was restored.

In addition, two research laboratories in two countries have found Tribulus to be very effective in improving sperm count, motility, and morphology, especially when combined with dietary and exercise changes.

The following herbs have been shown safe and effective when combined with Tribulus:

- Lack of libido: Damiana, Maca
- Male Sexual Dysfunction: Korean Ginseng, Saw Palmetto, Maca or Ashwaganda.
- Female Tonic for Reproductive Health: Shatavari, Maca

Depending on several factors, in many situations, nature would favour the conception of a boy (a male mammal in general) for reasons beyond the topic of this book. Clear examples of this fact are:

- The design of the Y-sperm, faster, to reach the egg first.
- Supplying the Y-sperm with a better "antenna" in detecting the emission from ovum's corona radiata, announcing the egg's location. X-sperms would experience repeated difficulties guessing in which fallopian tube the egg is waiting, resulting in more X-sperms entering the wrong tube.
- The Y-sperm has a unique ability to repair itself.
- The bodyguard type of sperm seems to favour protecting the Y-sperm. It is a fascinating scene under the microscope.
- When multiple Y & X sperms are attached to the egg, trying to penetrate it: If the woman is very healthy, the chemistry of egg surface area, is found to favour the Y-sperm . Women on a poor diet (and women in starving countries) conceive more girls than boys.

  Nature balances this situation by giving X-sperm a longer life span with a bigger body and a better survival rate. And nature would then favour the X-sperm when race survival is threatened (famine.)

  Woman's body is programmed, when both X & Y sperm are attacking the egg, competing for it, to favour the boy-producing sperm, only when the woman's brain tells the productive system that there is no "famine!"

  Fooling the brain, by skipping breakfast that day, may trigger the message to the ovum to favour a girl for race survival.

  Women who are "well-fed" and on a good diet, are more likely to bear boys (subject to all other discussed factors of course.)

Before you dismiss that, remember that many more facts we face today are based on programmes inserted in our system back in the jungle days.

Have you ever wondered why those who skip the morning meal or starve themselves, find it harder to lose weight?

Back in the jungle the man went into the forest at dawn, competing with the lions and tigers for a deer. He would come back in the morning hours and the woman or women would cook the "most important meal of the day" for them and for the children. There would be days when the hunt is unsuccessful and the family would spend the day looking for edible vegetation. When the morning hours come and go, and there is no "kill," the brain puts the body on starvation mode and lowers metabolism rate so than any stored fat would last as long as possible. Today, with a fridge full of food, you skip breakfast thinking that it would help you lose weight, when, in fact, it makes it harder on you to lose weight because the brain programme is on a starvation - low metabolism mode.

Another example. Why would a man sitting on a leather chair, in an air-conditioned office, suffer a heart attack under work-related stress? Back in the jungle we walked and ran barefoot in a forest where there are plenty of sharp rocks and broken branches. When a man was under stress it was associated with one danger, running away from a tiger or a lion. To prevent him from bleeding easily to death – if he suffered a cut by a rock or a tree branch, the brain sends a signal (a chemical compound) causing his blood to thicken, so that he would not bleed easily.

Now on a couch and in an air-conditioned office or at home, with sudden disturbing news, the brain interprets that fear and stress as being chased by a tiger, and thousands of years later the brain still sends the same "order" causing the blood to thicken – but now the man is not running, and his heart isn't synchronized with the same muscular activity when running in the jungle, and in an attempt to shift gear from park - on a leather chair, (when in fact, the brain assumes he is running away from the beast) – to 5th speed, it jams.

Another example, though I hate using this one because it not "politically" correct or a "smart" example to use when the audience is women. Man, at the beginning was not designed to be monogamous. All mammals are not. For social reasons, we humans at this stage, in many areas on earth, decided to go against man's inherited nature and to become monogamous. Of course it is the civilized and the romantic thing to do, and I support it.

At the beginning, man was faced with constant loss of his offspring to wild life, from wolves to lions, and to disease where there were no antibiotics or hospitals. Survival thus required a supportive number of children. Hence, he was programmed to "spread his seeds," to impregnate as many females as possible and produce as many children as he could. The woman on the other hand, sought the tough, rough, physically strong man who could protect her from the beast, i.e. she wanted the tough guy, not the nice guy - (and statistics still show that women married to successful white-collar men, in most cheating cases, they favour blue-collar men.)

Where am I going with this? Well today, a man has just had sex with his wife, he may be able to do it again to a limit; in most cases men are "finished" and are unable to "perform" over and over.

Back in the jungle the brain would send a signal (all hormones or enzymes are chemical compounds) ordering the body to turn the sex machine off, to save his "seeds" in order to be able to "service" another female, and to produce more children for survival.

The brain has not deleted this programme either. The same man today who has just finished making love to his wife, would feel he couldn't do it again (with his wife, right then) Yet, if another female, with appealing hips (able to carry children) in high heels and a short skirt walked in right there, he would be able to perform again. His brain would send that "order" to impregnate another female for race survival. This is why a man with a beautiful wife would still stare and to turn his head to "check out" another woman's hips, legs, or breasts. It's an instinctive response masked today by the pressure of the expected civilized or appropriate behaviour.

In several recent experiments, several male mammals (rats and rabbits) were introduced to a female. The male "serviced" her and lost interest. When another female was introduced he was aroused and again, performed and lost interest. A third female was introduced, and the result was typical … in one experiment the number of females who were "serviced" by one male in one session reached 40 before the male fainted out of exhaustion.

So you see, it is not hard to believe that the woman's brain would influence the ovum's favoritism of a male conception when it's a feast, and a girl conception when it's a famine.

A historical study in South America linked famine with giving birth to girls. It's nature's logic of saying we need girls for race survival, and we need only one man to impregnate many women.

A more recent study which links higher-energy intake around conception to the birth of sons provides the first explanation of why the number of boy babies is in decline in the west, suggesting it is the result of women consuming low-fat foods and skipping breakfast, among other things.

The research shows a higher calorie intake around the time of conception can shift the odds of having a son from 5 to 8 boys in every 10 births. The effect was, within that margin of measurement, women who had sons were more likely to have eaten a higher quantity and wider range of nutrients including potassium, calcium, and vitamins C, E, and B12.

In other words, women who want a son should eat a generous bowl of cereal for breakfast, munch bananas, use more salt and boost their overall daily calories by 500 calories - the equivalent of a meal - the study concluded.

The phenomenon of linking lean times with daughters has been most extensively studied in insects, but is also seen in horses, cows and some species of deer.

Females on the other hand, reproduce more consistently. It is known from two university research papers that high glucose levels encourage the development of male embryos while inhibiting female embryos. In humans, skipping breakfast depresses glucose levels and so may be interpreted by the body as indicating poor environmental conditions and low food availability.

Another research paper stated: "This is an interesting result that is consistent with what appears to be going on with some animals such as red deer."

Considering the "programmes" that have been inserted into our brain since the beginning, it makes sense to assume that the brain influences the female chemistry in favour of a girl, when the brain believes that the body is "sick" or weak. This is because many hereditary diseases the woman could carry, she can only pass them to a boy, not to a girl. Diabetes, muscular dystrophy, hemophilia and many enzyme deficiency disorders would pass on to a son, not to a daughter. Hence, if the brain feels the body's fitness or condition is "sickly" it may try to protect the next generation by favouring the conception of a girl.

We maintain of course that the diet information is not by any means sufficient to depend on it as a sole method to select the gender of your baby. It's only a secondary, supportive element; a small part of the larger, complete picture. It's much more effective of course, to aim for eliminating one sperm or another entirely, and to secure – practically 100% – the success of gender selection, as I'll explain in the following chapter.

For mammals other than humans – mainly Great Danes, I have, through Optimum Green Laboratories, developed a supplement of nutrients, vitamins, hormone enticement compounds, for the purpose of triggering the female's ovulation, i.e. inducing her estrus in order to perform the AI impregnation in late winter, and avoid whelping in the harsh summer or harsh winder months.

We have also introduced a formula to enrich the male's sperm production. These formulas are currently in the red-tape stage.

Nevertheless, a plethora of qualifying research papers are in support of the direction and the theme of these formulas. As I selectively presented, diet can play a role in improving the general health of ovulation aspects, and a secondary supportive element in assisting either the "Y" or the "X" sperm to penetrate the ovum.

Synopsis:

- Consume a great amount of calcium and magnesium and very little potassium and sodium when trying for a girl.

- Increasing sodium and potassium levels in the body enhances the fertilization of the egg by a Y-sperm. The woman's diet should be rich with items such as bananas, and must eliminate items such as milk chocolate, cheese and commercial desserts.

- Low levels of sodium ions diminish receptor activity. Decreasing potassium and increasing calcium and magnesium attracts K & Mg ions to these receptors. This combination turns the table in favour of the X-sperm.

- Eat plenty of boiled oysters with no salt for months prior to conception when trying for a girl.

- Consume a great amount of potassium and sodium and very little calcium and magnesium when trying for a boy.

- Eat plenty of fish and red meats for a few months prior to conception when trying for a boy. Another source of high potassium is table wine. (note: no drinking, in fact the entire diet becomes unnecessary, once you're pregnant) – Also for a boy, cowpeas, especially if immature, are very high with potassium.

- Desserts with milk-base such as custards and puddings are high on calcium, low on potassium and sodium, favouring a girl.

- A drink high in caffeine such as tea, coffee or chocolate by the man prior to intercourse favours the Y-sperm.

- For a boy, eat raw fruits and veggies, not canned or cooked or frozen. Also eat steaks and roasts. Avoid milk and milk products like puddings and custards.

- For a girl, eat very little salt. Minimize pop. Avoid baking soda or baking powder in your cooking. Avoid alcohol, instant cocoa, ice cream, commercial chocolate syrup and similar products that contain sodium or potassium. Take a 50 mg vitamin C tablet daily.

- The woman must go off the diet immediately after getting pregnant. The pregnancy diet is another subject, and you would need to compensate for the low side of the pre-pregnancy diet, and to attend to the needs of the pregnancy.

Notes:

A word about OGL supplements.

www.optimumgreen.com

Note: ***You may skip this chapter****, or read it later as a bonus material. It is relevant to the concept of hormonal physiological control.*

Optimum Green Laboratories operates five research and commercial laboratories. Each branch is operated by a doctor-chemist as its manager - head scientist. A panel of seven chemists – MDs form the technical committee. OGL was founded in Ontario in the early 1990s by two chemists, one of whom is the author of this book, its present owner and chief scientist.

OGL has been researching what was described as an ingenious idea, introduced by Dr. Gouda several years ago:

"mimicking the brain response to manipulate the body's reaction in the same safe, natural way the body reacts to its hormones and enzymes."

The research, which has been led and directed by Dr. Gouda personally, has finally produced what promises to be a revolutionary invention in medicinal chemistry.

Four main formulas are making their way through the red tape. They deal with:

- Erectile deficiency treatment from a new perspective, totally different from the approach taken by common present drugs that are based on the concept of inhibiting an enzyme reaction and producing sufficient relaxed dilation by releasing NO "nitrous oxide" to allow blood into the penis. This current approach comes with negative side-effects, including loss of vision even stroke.

The new approach is based on a serious of natural chemical equations that results in duplicating the brain's signal (or order) impeded in the old programme we have inherited. I explained in the preceding pages how the male mammal is programmed to be re-aroused to fertilize more females. That chemical response is mimicked via a specific organic approach.

This approach is designed to naturally trigger the body to, for example, reverse testosterone deterioration and to mimic the production of dopamine and testosterone. As a result, the brain tells the body "you are young" and it starts to develop better masculinity. It also has a positive side-effect, prostate health improvement. The author himself witnessed a significant drop in his PSA from an alarming life-threatening level, to, in one year, a 25 years old's prostate condition.

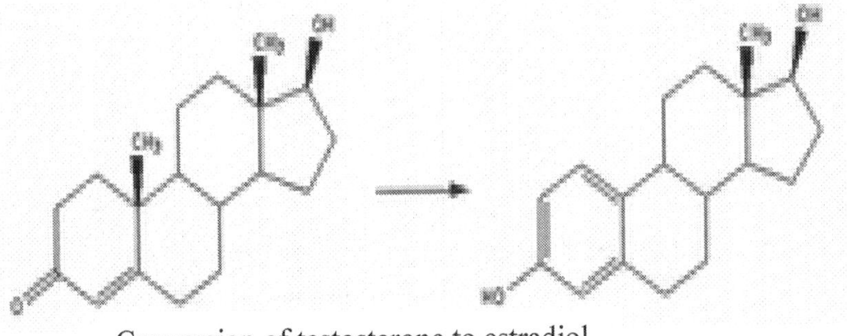

Conversion of testosterone to estradiol

- Secondary formulas are designed for male mammals to enrich semen quantity and sperm health, and, there is another formula to enrich the Y-sperm in particular.
- Ovulation stimulation and ovulation health improvement. This was complimented by a secondary formula designed for mammals other than humans to induce the female's estrus.
- Using the same approach, a formula was introduced to mimic hormonal reaction in the body, to manipulate cell

deterioration that leads to aging appearance. This has shown incredible potential, especially in women.

- And finally, mimicking the body's hormonal reactions to improve memory retention in the same natural way hormones behave in our younger years.

These are obviously going to be major news, and you are here given the pre press-release review. The protective approach to it – beyond the basic patent procedure, and the investors' and marketing agent demands to delay publishing its details at this stage, until a secured assurance of marketing protection is not a concern - is obviously understandable.

I wish to clarify some misconceptions first:

- "Natural" does not mean pulling an herb out of the soil and eating it. That could be fatal. The same herb that may have positive medicinal properties, may also have more serious negative properties. It's the chemist's role to extract the "good" away from the "bad."

- Everything is "chemical." The Chemists'Association's motto is "what in the world isn't chemistry?" If you say, "Di-hydrogen monoxide," it may sound technical, chemical, not natural - when it's the definition of water, the most natural material.

It's just whether it's a good chemical or a bad chemical. Oxygen – which we breathe in all the time, just one more atom, would change the safe water H_2O to H_2O_2 a bleach-like hydrogen peroxide.

Our body, water, light, food … everything is a chemical compound. Our body's cells, our blood, even our feelings such as love or fear ... the hormones and enzymes that trigger a certain feeling are chemical compounds.

Testosterone $C_{19}H_{28}O_2$ the chemical of masculinity is just another compound. All "signals" or "orders" from the brain are compounds. We chemists found a way to manipulate and mimic many of these compounds.

In a slightly technical tone, here is an example of how we chemists can utilize basic physiological facts to produce compounds to manipulate brain control of our favour.

There are several simple, natural chemical reactions and chemical compounds that the body controls, which we can duplicate and manipulate to our advantage.

Free testosterone (T) is transported into the cytoplasm of target tissue cells where it can bind to the androgen receptor or can be reduced to 5α-dihydrotestosterone (DHT) by the cytoplasmic enzyme 5-alpha reductase. DHT binds to the same androgen receptor even more strongly than testosterone so that its androgenic potency is about 5 times that of T. The T-receptor or DHT-receptor complex undergoes a structural change that allows it to move into the cell nucleus and bind directly to specific nucleotide sequences of the chromosomal DNA. The areas of binding are called hormone response elements (HREs), and influence transcriptional activity of certain genes, producing the androgen effects.

The ratio of the levels of circulating estradiol or testosterone, the "free" and the bound fractions, and free androgen index expresses the ratio of testosterone to the sex hormone-binding globulin and can be used to summarize the activity of free testosterone.

A number of synthetic analogs of testosterone have been developed with improved bioavailability and metabolic half-life relative to testosterone.

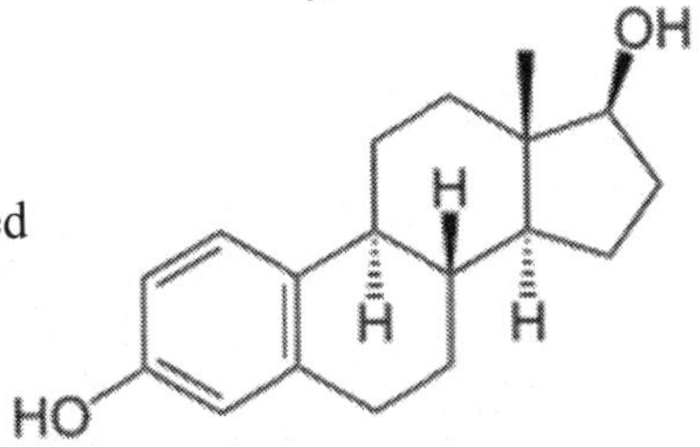

Many of these analogs have an alkyl group introduced at the C-17 position in order to prevent conjugation and hence improve oral bioavailability. These are the so-called “17-aa” (17-alkyl androgen) family of androgens such as fluoxymesterone and methyltestosterone.

.

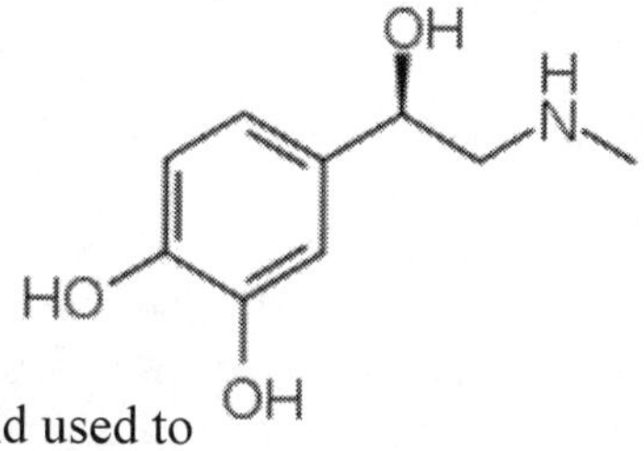

L'arginine is an amino acid used to enhance the effect of nitric oxide; an essential compound for normal sexual performance. It naturally relaxes the muscles surrounding the blood vessels that go to the penis which results in strong blood flow to the penis and thus obtains and maintains an erection.

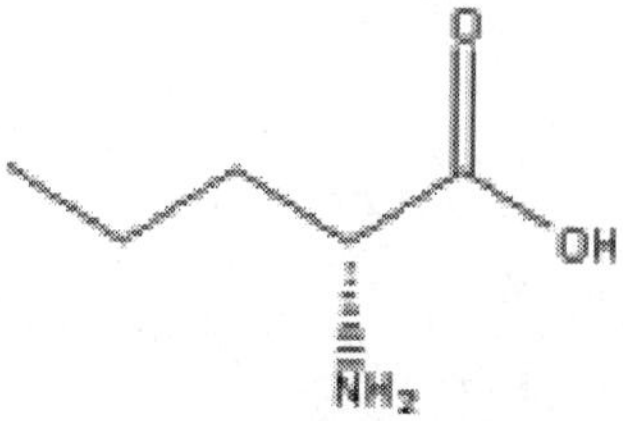

Bioactive Glygopeptide compounds are proven to increase free testosterone and decrease SHBG levels in human clinical trials. Thus, you would achieve a healthy erection and a healthy sex drive.

In the lab, we can condition and maximize sprotodioscin, a saponin and other compounds proven to improving libido and sperm motility. The activity or excitement of such a compound triggers the brain to send that magical order to the penis. This is well-illustrated in Dr. Gouda's technical paper.

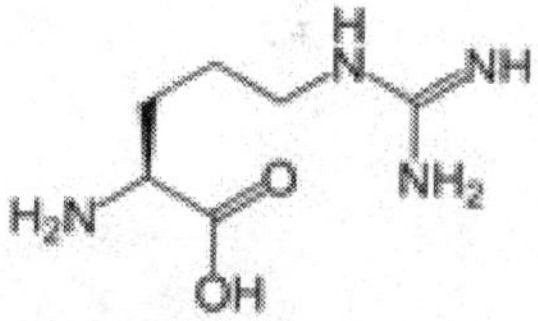

L-Arginine HCI manipulation is another example of what OGL did to mimic the body's natural response, simply duplicating natural chemical equations.

Hormones are very potent substances, i.e. a very small amount of a hormone may have profound effects on metabolic processes. As a result, hormone secretion must be regulated within very narrow limits in order to maintain homeostasis in the body. Several hormones are controlled by some form of a negative feedback mechanism. In such a case, a gland is sensitive to the concentration of a substance that it regulates. A negative feedback system causes a reversal of increases and decreases in body conditions in order to maintain a state of homeostasis. Some endocrine glands secrete hormones in response to other "tropic" hormones.

OGL's approach is designed around the concept of positive manipulation or hormonal action control with the formulas (or the chemical equations we naturally inherited) - and thus we duplicate the natural human reaction.

This achievement is a product of the alloy of "chemist" and "MD" amalgamation. Most people don't realize that typical MDs are not chemists, they know little about medication, and they spend their lives around physiological symptoms, treating them (and getting credit and recognition) based on what chemists have invented, and the drug information chemists send them.

I had the option to pursue a GP doctor career, and I am glad I changed gear into pharmaceutical research and analytical chemistry. The point is, this book is supported by knowledge beyond the mere MD's radius. And the awaited paper on hormonal chemistry manipulation is supportive research that will be published by the time the next issue of this book is reprinted – subject to the investors securing satisfactory commercial protection measures.

"The approach of hormone manipulation invention by Dr. P.H.R. Gouda, subject to official product registration – qualifies him for the noble prize in science nomination" – the EPL chief scientist was quoted, expressing EPL's intent to initiate such nomination recommendation in time.

The innovative approach of chemical manipulation of hormones has an incredible potential. Dr. Gouda's marketing agent – representative, is in the process of negotiations with pharmaceutical agents to materialize the use of such invention – subject to red tape and investors' commercial formalities.

OGL has produced several effective formulas.

The group "A" products that are relevant to this topic are:

- Supplements for women diet prior to conception – boy.
- Supplements for women diet prior to conception – girl.
- Supplements for general female fertility health.
- Supplements for general male fertility health.

The four formulas of group "B" are:

- E.D. correction via hormonal manipulation, and the associated positive side effects of prostate health improvement.
- Anti-aging – hormonal manipulation approach.
- Memory-retention improvement and prevention of dementia cases.
- The detailed study into customized, individualized vitamins and minerals profile specifications.

… These "A" & "B" products should be available in European pharmacies soon, and in the North American market hopefully not much later.

Notes

A.I.

Artificial Insemination.

The success of gender selection is, in my opinion, divided into five categories: A, B, C, D & E – as illustrated below.

A: represents the effectiveness and success range of the basic exercises designed around the use of sperm characteristics: e.g. Y-chromosome sperm is faster and short-lived or less resistant to acidity while the X-chromosome sperm is slow but more hardy. And, as such, controlling acidity and distance would favour one over the other, which makes timing of intercourse, relative to ovulation, a critical factor. Well, depending on how much and how well you cover this area (illustrated as 3 stages or degrees of successful adoption under "A") you would on average be expected to succeed within the "A" area, with a potential to reach "A2" peak.

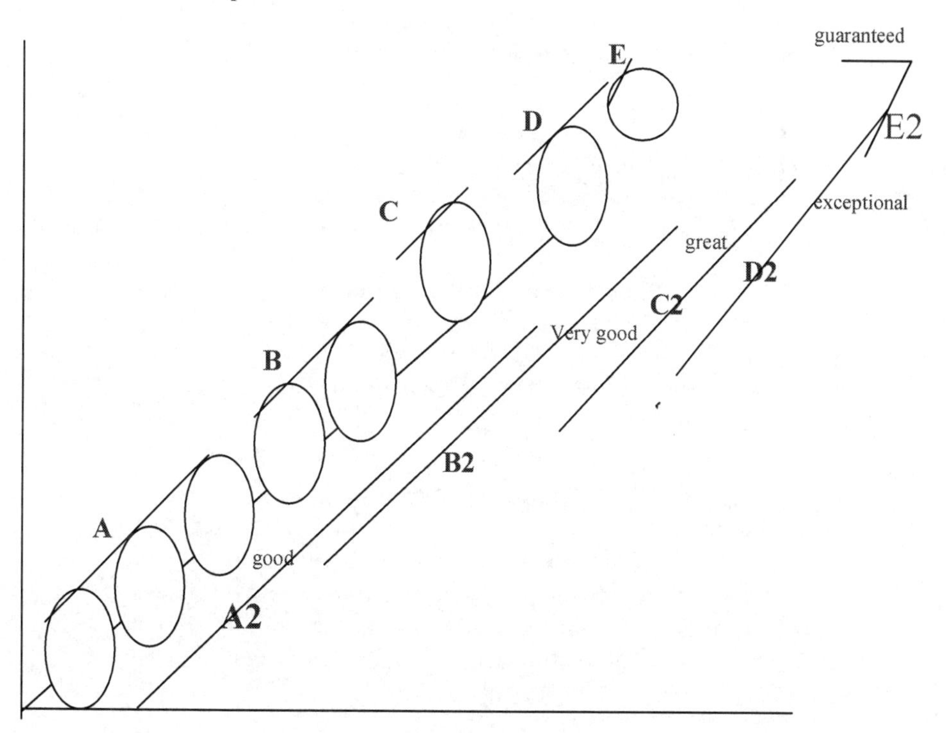

The second degree of success, "B," is achieved when in addition to, or in combination with, the approaches covered by "A". You would also adopt the full supportive supplementary diet.

Stage “B” covers incredible success potential with no practical need past this stage. It is reasonable to assume here that you would have minimized potential failure beyond reasonable doubt. Stages C and beyond are aiming for the absolute-no-doubt goal.

Stage “**C**” is when you employ conditioned artificial insemination.

Stage “D” employs sorted sperm, CAI.

And stage “E” involves IVF insemination. I’ll explain.

You’ll see that “E” is not necessary. Reaching the highest potential of “D” achieves near that 99+ goal.

First, let me talk about what I call it:

CAI

Conditioned Artificial Insemination.

I must state that without CAI (but with intelligent implementation of the basics along with a good diet), you will still get a good ≥80% odds in your favour. Yes, with just applying stages “A” & “B” approaches of ovulation determination, vaginal secretion conditioning, controlling depth and quantity of sperm deposit, and diet with all the tips introduced prior to this chapter, you would be increasing the odds in your favour significantly. Surely it would be worth it to try just that – without getting into AI. Increasing your success rate to well over 80% is incredible.

One comment I must re-emphasize, is my disagreement with the old books on the market advising you to have intercourse a day, or about 24 hours prior to ovulation when trying for a boy.

Again, this is too long a waiting period for the Y-sperm. When using both an ovulation detection kit several times a day during the questionable couple of days, along with temperature monitoring, and thus having achieved an accurate determination of ovulation supported by charts of the previous 3 months of homework, you would be able to have intercourse an hour or so just prior to ovulation.

However, we are now talking about a different degree of certainty. Almost absolute certainty. And that's what the next chapters will cover.

CAI takes us to a whole different level of success.

I totally disagree with the books that didn't appreciate A.I. No book however that we are aware of, has addressed CAI – which, according to our record, is a technique accredited to Dr. Gouda via his paper "C.A.I. on mammals for the purpose of sex selection."

The couple of doctors who wrote about A.I. not CAI, likely have missed something … done something wrong. They are absolutely wrong in believing that A.I. is not exceptionally effective, or as they believed, not as effective as natural intercourse.
They are dead wrong.

A.I. is by far more effective in achieving pregnancy than natural intercourse. In 10 years of experience, I have never had one single failure. I even have, casually as a friend, coached or advised couples on performing it at home. Which they did, without medical training other than a session with me on DOs & DON'Ts.

And, my other mammal experience in dealing with horses, Great Danes, and cattle never failed either, not once. Not only it was 100% successful impregnation, but it also utilized CAI, for specific sex selection.

This categorizes it, at least at a selective level, reasonably within the 100% success rate – and 10 years of wide mammal CAI application could hardly be called a small sample.
A.I. will do a perfect job.
Why?
You will be able to deposit the semen manually, as deep as you want, right by the cervix if you want a boy, all of it, and in an optimized or conditioned state. I'll explain what I mean by "conditioned."

I must ask you to review the chart of the female productive system again. Now, here is the simple tool you will need:

Collecting the semen sample is simple and needs no further comment here other than some basic precautions you can obtain from your GP.

Semen collection bags are available at specialty clinics and drug stores, or online. Maintaining sanitization is obviously a must.

The sample is then injected into the female.

There is, I must admit, more to it than that and I don't recommend you would just get a syringe, and an insertion tube and perform it. It takes a bit more knowledge and precautions, and you would need a good consultation session with your doctor – if, for any reason your doctor would not do it for you.

Conditioned A.I. allows you to:

- Excite, enrich, and stimulate the sperm – or weaken it, stress it or dilute it – depending on the objective. The technical paper designed for medical professionals will address this in detail.

- Use techniques to favour the selection of one sperm type over the other. I'll show you how.

- Condition the semen sample further in a test tube (or collection bag) prior to deposit, for an optimum result.

I have utilized my analytical laboratory bench experience and sample-prep apparatus and instrumentation to achieve specific conditioning – as I'll explain. However, a great deal of the treatment can be achieved with minimal training or equipment.

And your family doctor, any dedicated GP, or a friend who has attended medical school, or any friend who is a chemist or a pharmacist - or a qualified a nurse - will be able to help you with the next step of the procedure. (Your GP may not assist you if the health care system in your local area will not cover it – or if liability concerns or other legalities are factors.)

A final note, while baby sex selection does not necessitate A.I. or CAI. (they ares a bonus) obviously sex selection in animals requires A.I. and in fact, CAI and even sperm sorting.

Notes:

Conditioning a collected sperm sample for AI involves two stages.

The first stage involves these options:

1- Dilution – reduction to lower sperm count.

2- Radiation and/or heat treatment (and other stress options, performed in the collection bag) to lower Y-sperm count. This requires specific skills in order to not harm all sperm. My experimentation data is part of the technical study paper. In fact a hot steam bath will help you reduce Y-sperm count, and that is a kind of conditioning, you may say, with no need for A.I.

3- Addition of nutrients to stimulate sperm performance. I have developed a glucose-based formula that will be explained in the technical paper. It was found to be very effective in mammal sex selection via CAI.

4- Stress and delay time to favour X-sperm performance.

Then,

The next stage takes us to the ultimate category.

The actual sperm separation. And that's a different topic all together.

CAI employed all the preceding techniques back in the 1990s, the results of which were published by Dr. Gouda in his report "CAI & mammal gender manipulation – with and without sperm sorting."

Meaning that, what Dr. Gouda has experimented with on mammals, involved collecting the semen sample, and prior to inserting it into the female, he either only "conditioned" it and deposited it, or conditioned it and then performed sperm sorting. By performing such "conditioning" he had repeatedly, over a decade, been able to achieve 100% gender selection success.

The same CAI mammal application approaches introduced by Dr. Gouda in his paper can be adopted when approaching the mammal, humans.

Obviously for legal reasons- as well as for the need for approximately 200 pages of carefully addressed instructions, and considering that such approach would be too technical for a layman to perform without supervision, the technical part is kept separate and apart from this book. It will be published by Dr. Gouda after his planned clinical study, and subject to volunteer response and other involved formalities.

It remains however that calculated and careful heat application, stress and pH control, or treatment of a collected sample have shown an absolute success in the CAI practice in mammals, and yet they involve rather simple procedures.

A.I. also allows us to minimize the influence of ovulation time determination. Knowing that ovulation has already occurred, depositing the sperm after ovulation becomes "not an issue." The reason is the ability to bypass the harshest obstacle the sperms face, the initial vaginal acidity – this allows us to deposit the sperm deep near the cervix.

The conclusion is, A.I. is simple, and with minimal technical help, it can offer you an effective fertilization tool when sex selection is a factor.

Fertilization via sperm separation can be divided into two categories:

- Separation of sperm followed by administering A.I. - i.e. the sperm is inserted into the female.

- Separation of sperm, and test tube fertilization using an egg extracted from the female and reinserted into her womb. This is known as IVF with PGD (In Vitro Fertilization with Pre-implantation Genetic Diagnosis)

The IVF method has a success rate of nearly 100 percent. It is, however, expensive, must be done at a professional clinic, and is much more physically intrusive for a woman compared to basic sperm separation AI. Neither method will cause any harm to the developing baby.

Sperm separation - (also known as sperm sorting) is the next step up from CAI.

The advantage of CAI is the fact that with minimal training, one would be able to perform the sperm conditioning, followed by AI, effectively. Sperm sorting however, depending on the technique, may require an experienced professional and specific laboratory apparatus and instrumentation.

Centrifugation is an effective method to separate the X and Y-sperm.
It is based on controlled spinning, causing particles in a sample to become sorted into layers according to their density.

Sorting separates the more dense X-sperm from the lighter Y-sperm. During sperm sorting the sperm cells are separated from the seminal fluid. The sperms are concentrated into healthy and motile sperm, abnormal sperm is filtered out, and finally, the X and Y-sperm are separated.

This technique is the SOP in the IVF and IUI procedures, but it doesn't have to be limited to them. Any proven sperm sorting method can be used with A.I. After the sperm is sorted, the sample will be inseminated into the woman – or the female mammal, which is an "in-office" technique. The most well-known method of sperm sorting is the Ericsson Method.

Sperm sorting, in this case, utilizes the flow cytometry technique to analyze and separate spermatozoa.

During the early to mid 1980s, Dr. G. Spaulding was the first to sort viable whole human and animal spermatozoa using a flow cytometer. He utilized the sorted motile rabbit sperm for artificial insemination.

Thirty years ago, sperm separation took a different turn. The approach utilized the DNA difference between the X and Y chromosomes. Prior to flow cytometric sorting, semen was labeled with a fluorescent dye compound that binded to the DNA of each spermatozoon. As the X chromosome is larger, thus with more DNA than the Y chromosome, the "female" (X-chromosome bearing) spermatozoa will absorb a greater amount of dye than its male (Y-chromosome bearing) counterpart.

Accordingly, exposure to UV light during flow cytometry makes the X spermatozoa fluoresce appear brighter than Y-spermatozoa. And, as the spermatozoa pass through the flow cytometer, they are separated.

IVF with PGD (In Vitro Fertilization with Preimplantation Genetic Diagnosis) procedure is accepted as a 100% successful method. IVF (meaning fertilization outside the body in a test tube) - is not commonly necessary.

The separation of spermatozoa from the seminal plasma is required to prepare the spermatozoa for intrauterine insemination, in-vitro fertilization, and sex selection. The study looked at the effects of sperm preparation techniques on the functional integrity of the sperm membrane as measured by the hypo-osmotic swelling test (HOS).

Semen analysis and HOS performed on each aliquot, produced mean and standard deviation for the HOS of 72.9 ± 8.5%
71.2 ± 13.1 after Percoll
75.2 ± 15.1 after swim-up
and 62.4 ± 14.5 after Sephadex.

Analysis of variance showed that the mean HOS score was the same after Percoll and swim-up as it was initially. It was considerably less following preparation with Sephadex.

There was also a higher proportion of abnormal semen specimens (HOS <50%) after preparation with Sephadex than after the other preparation methods. The recommendation here was to use the HOS test as part of a screening panel for sperm separation. As I indicated, I'll address all that in the detailed technical paper.

I disagree with the common concept of the association of sperm sorting and test tube fertilization. IVF is not a must do.

Sperm separation can be performed, and the selected sperm is then deposited via CAI into the female, not necessarily via IVF.

The fact that a smaller sperm sample will be selected is compensated by the fact that technology allows us now to deposit the sperm deep next to the ovum. Not to mention that sperm separation via different methods allows us to obtain a good-sized sample. In fact, my personal examination of the theory – and the possible theoretical complications – of this approach, led me to vote against PGD at this stage of development despite its secured success as far as gender impregnation. However, this is again a technical area I'll leave its coverage for my detailed and technical paper.

I'll briefly, however, make a few very short statements.

The scientific community, in general, considers the real-time sperm separation technique to be the most effective method so far available in isolating high-quality sperm samples.

A significant issue I'll address in my detailed paper deals with the effect of swim-up, Percoll, and Sephadex sperm separation methods on the hypo-osmotic swelling test. Your doctor can brief you on all this – as getting into details of issues beyond your role at home – in terms of your role in sex selection, is obviously unnecessary here.

Depending on the deposit procedure, sperm separation – a technical procedure, can be simplified to offer a valid option.

This is exactly what I have done for a decade with a 100% success rate:

1- Female diet & ovulation health. A unique supplementary programme. My own formulas.

2- Male diet and the other specific conditioning elements. Again, my own OGC formulas.

3- Ovulation date relative to intercourse or to A.I.

4- The basic factors: vaginal pH control, depth of deposit, quantity of deposit.

5- C.A.I. which involved semen stress, heat treatment, dilution … and simple "layer technique" separation prior to A.I.

6- CAI involving sample treatment in terms of enrichment, dilution, applied stress, or sperm separation. This portion of the procedure is not necessary for a very good success rate, and is not a do-it-yourself at home practice without professional guidance (hence they are not addressed here in illustrative details) - however, this is a bonus that would improve the near-perfect odds. My personal mammal CAI experience is testimonial to this fact.

These steps do not have to involve sperm separation, though applying simple layer separation technique would be a big additional bonus to CAI. The combination is too powerful for any practical chance of failure.

In conclusion,

With diet, accurate ovulation date determination, the correct deposit depth and quantity of sperm, female vaginal conditioning, and the few simple facts and factors addressed by the first part of this book, without A.I. or CAI, and certainly without sperm-sorting, you can, with great certainty, select the sex of your baby.

Sperm separation for CAI

Sperm separation seems to be addressed by many doctors (who likely never practiced it once) as a foreign procedure to be feared.

There are several sperm separation techniques that can be easily and safely performed with minimal technical knowledge or supervision.

In the 1960s an Egyptian doctor separated the sperms by placing them in a thick solution where sperm movement would be harder, and the Y sperm swam faster to the bottom. Separating the top and bottom layers which were monitored under the microscope, rendered a very high concentration of one sperm in each layer.

I have applied these techniques at home, performing CAI on other mammals. In fact, only just three months ago, I formulated a jelly-type glucose based solution in a tube, to separate sperm in performing CAI on Great Danes. These are the same techniques that were performed on humans in several European studies with up to a 97% success rate in 5 documented cases. The separation techniques are again based on the simple fact associated with the characteristics of the Y & X sperm.

In a special apparatus, we can easily, deposit the total sample at one end of a special laboratory column, in a specific albumin medium while monitoring sperm movement. We then obtain the isolated faster Y- sperm in the advanced layer of the sample, leaving the X-sperms behind. This has been done for decades by doctors from Japan, Egypt, Germany and Holland.

Surely you may have a few X-sperms in the collected Y- sample, but that's exactly it, few. Logic, or science dictates that the odds or probabilities are by far in favour of the faster majority Y-sperm. This is a combination that in any practical term would guarantee the success of a Y-sperm if the Y-layer was used, or the success of X-sperm if the X-layer was used.

Another separation method utilizes an approach every bench chemist is very familiar with, in fact, every lab technician is. Spinning the sample in a test tube – based apparatus to isolate layers based on specific gravity or density – which differs between the heavier X-sperm and the lighter Y-sperm.

This is a very simple procedure that, although of course, you may still have a "few" undesired chromosomes in the sample, yet, with the application of the complete procedure, this minority does not stand a practical chance. Such (almost perfect) sperm separation is good enough to raise the success rate from the 70s% to the high 90s% rate.

Laboratory centrifuges and other clinical centrifuge equipment are simple effective tools in sperm separation for A.I. procedure.

I have separated the two sperm types at a small lab, and in fact at home when I inseminated animals. The sperm sorting technique was as simple as placement of the semen in a solution, allowing a 30-second race in a tube.

Another conditioning was as simple as heat application to stress the Y-sperm. Both personal techniques were 100% successful.

Think about it! If just a good adoption of the diet approach alone has achieved an 88% success rate - and the simple utilization of sperm characteristics such as speed and longevity has achieved 70% - and sperm separation could give you a 95% successful sorting (i.e. 5% only of the undesired sperm type) - all other achievable percentages by other approaches would now have their failure margin within that 5% only. Meaning a 30 percent failure, is 30% of 5. You have already secured a 95% success with separation. Add all these together, addressing the topic from every angle and utilizing every advantages, will add up to ≥99%. Practically 100% success rate, which is my own personal record.

A closing note

I must remind the reader that:

- No material here should be considered without considering your own personal health profile based on the advice of your doctor. Whenever I had any doubt that I was about to present material too technical for the layman, and thus might constitute "playing with fire," I refrained, summarized, and I offered either a way around it, or emphasized the need to work with your doctor.

- The diet presentation refers to pre-pregnancy. Once pregnant, the diet must be adjusted with one objective in mind, a healthy baby and a healthy mother.

- Nothing in life, except death, is absolutely guaranteed. Despite the (practically, or almost 100%) success expectation - if you satisfy all aspects of the concept, and despite the actual 100% success I have achieved thus far, you must consider the possibility that you might be that rare 0.1% or even 0.01% case exception. Do not attempt pregnancy unless both parents are 100% certain that they will love and care for that baby regardless of gender.

Having said that, I have no doubt that the official complete technical paper to be submitted for publishing with the medical journals, which will be based on a large pool of volunteers, will confirm this conclusion and will announce to the medical society, and to the community at large, the good news that many have long awaited.

To conclude, here is a summarized list:

A boy list. And a girl list.

A boy list:

* The man must refrain from sex for a few days.

* The man must avoid testicular exposure to heat and stress.

* The man must go on a diet rich with zinc, iron, multi vitamin and minerals and a high-calorie intake diet. See diet section. Also keep in mind details such as the need to take sufficient vitamin A if you are taking extra zinc.

* For the man: a dose of caffeine prior to intercourse (e.g. coffee.)

* Deep penetration and a strong ejaculation. See sex position.

* The woman's diet for 2 months prior to pregnancy must be very rich with potassium and salt, and very low on calcium and magnesium.

* Woman's diet must be of high-calorie intake, especially during the week of intercourse, and special attention must be given to the day of intercourse with several meals starting early in the morning.

* The woman must eat plenty of fish, red meat, bananas, salt, table wine. Avoid milk chocolate, cheese, and commercial desserts.

*Ovulation date determination. Intercourse very close to ovulation.

* Use of alkaline vaginal wash prior to intercourse as explained.

* The woman should experience orgasmic excitement, preferably before man's ejaculation.

CAI, with or without sperm separation, would be the cherry on top that would take you from "very good" and "great" odds, to near that 100% guarantee. OGL supplementary formulas are recommended. www.optimumgreen.com

A girl list

* The woman's diet for two months must be very rich with calcium and magnesium and very low on potassium and sodium (almost no salt, and very little or no bananas or red meat.)

* The woman: eat plenty of boiled oysters with no salt. Also eat plenty of desserts with a milk-base such as custard and pudding. Avoid pop, alcohol, instant cocoa, ice cream, chocolate syrup.

* Take a 50 mg vitamin C tablet daily.

* Do not maintain a high calorie-intake diet. Do not have breakfast for 3 days prior to intercourse and eat 2 meals of low calorie intake for the days prior to intercourse.

* Do not enjoy sex this time. Think of something terrible he did that irritated you.

* Use an acidic vaginal wash prior to intercourse (see earlier material.)

* The man must masturbate or have protected sex daily for a few days, including the morning of or the night before the attempted pregnancy.

* The man should wear polyester tight shorts or underwear for a few days.

* Keep the penetration shallow, not too deep, and upon the start of ejaculation, use your imagination to distract your excitement and to minimize sperm deposit.

* Time intercourse 2-3 days before ovulation.

CAI, with or without sperm separation, would be the cherry on top that would take you from "very good" and "great" odds, to near that 100% guarantee. OGL supplementary formulas are recommended. www.optimumgreen.com

This book was first published by "Great Canadian Books" under ISBN number: 978-1-927070-02-4 as a limited, conditional edition.

It was filed with The Canadian Intellectual Property Office.
Certificate of registration of copyright number: 1096166

And it has been registered with The Library and Archives Canada Cataloguing in Publication number:
QP279.G69 2012 613.9'4 C2012-903590-4

This 2nd edition was published in 2012 by iUniverse. It was registered with the Canadian Archives Library In Publication, and the USA Library of Congress.

All inquiries must be filed under the currently valid ISBN issued for this edition by Iuniverse Inc.

The book has also been made available in 2012 at Chapters – Coles – Indigo and at Barnes & Noble bookstores, as well as via their sites:
chapters.indigo.ca
bn.com

In the absence of shelf-copies in any particular Chapters – Coles - Indigo or Barnes & Noble bookstore, the book can still be ordered at the store via their database "search & order" PCs located inside the Chapters, Coles & Barnes & Noble bookstore.

The scientist

The author is a scholar with an elite academic status in both science and literature.
He is a pharmaceutical, analytical research chemist. After first attending medical school, and realizing that becoming a GP wasn't his calling, as he poetically put it, he just could not resist the laboratory scent, the test-tube feel and his passion for analytical chemistry, he switched into pharmaceutical chemical research.

His graduate studies at the Masters and the Ph.D. levels were in analytical and pharmaceutical – medicinal chemistry in both research and development – as well as pioneer papers in environmental chemistry producing two university textbooks and several analytical SOPs that were named after him.

The author, who was facetiously called a "professional student" - has also led a different scholastic life. His extensive academic background presents years of graduate "hobby" studies in sociology, theatre, medieval theology and psychology that were crowned with a second doctorate in literature. He is a published poet.

He is also a renowned athlete with international football "soccer" participation as a referee and instructor. And, he obtained a Mensa membership back in 1992.

The author is a single father to one child. Dr. Gouda and his only son reside on Vancouver Island, British Columbia, Canada, and he is pursuing entrepreneurship as an owner of a hospitality business and a commercial analytical laboratory.

The author can be reached at
www.goudabooks.com
chemist@doctor.com

Dr. Paul H. Ramses Gouda

Other books & papers by the author.

* Hydrargyrum – Hg analysis from diphenylthiocarbozone to cold vapor.

* Metalloid ultra-trace analysis of As, Se, Sb by hydride generation.

* Comparative background interference: Atomic absorption VS inductive coupled plasma. Environmental research.

* Pharmaceutical compounds, toxicity.

* Chemical manipulation of neurochemical compounds for medical research & development.

* Physiology manipulation in the field of sex selection for both human conception and animal breeding applications.

The author has also published several books in literature.

Notes:

Made in the USA
Lexington, KY
13 January 2016